THE
RESTLESS UNIVERSE

MAX BORN

Figures by
DR. OTTO KOENIGSBERGER

Authorized Translation by
WINIFRED M. DEANS, M.A.(Cantab.), B.Sc.(Aberd.)

DOVER PUBLICATIONS, INC.
NEW YORK

Published in Canada by General Publishing Company, Ltd., 30 Lesmill Road, Don Mills, Toronto, Ontario.
Published in the United Kingdom by Constable and Company, Ltd., 10 Orange Street, London WC 2.

This Dover edition, first published in 1951, is an unabridged and corrected republication of the work originally published in 1936. This edition, which also contains a new Postscript by the author, is published by special arrangement with Harper and Brothers.

Standard Book Number: 486-20412-X

Library of Congress Catalog Card Number: 51-13192

Manufactured in the United States of America
Dover Publications, Inc.
180 Varick Street
New York, N. Y. 10014

To
My Son
GUSTAV

TO THE READER

You may have wondered why so many of the pictures in this book are so like one another. No doubt you have already found out how they work. I am going to try and tell you something about the restless universe and to let you see something of its inner secrets for yourselves! You will come across from time to time references to " Film No. so-and-so ", you then turn it up and work it. The pictures all run from the middle of the book outwards. For those in the first half of the book it is best to hold the book in your right hand and flick over the pages with the left thumb; for those in the second half, use the opposite hands. First run through each " film " quickly, then more slowly, and watch carefully exactly what happens.

M. B.

CONTENTS

CONTENTS

LIST OF PLATES

FILMS

Aber im stillen Gemach entwirft bedeutende Zirkel
Sinnend der Weise, beschleicht forschend den schaffenden
    Geist,
Prüft der Stoffe Gewalt, der Magnete Hassen und Lieben,
Folgt durch die Lüfte dem Klang, folgt durch den Aether
    dem Strahl,
Sucht das vertraute Gesetz in des Zufalls grausenden Wundern,
Sucht den ruhenden Pol in der Erscheinungen Flucht.

Schiller, Der Spaziergang.

But in his quiet chamber the pondering sage describes
Magical circles, and steals e'en on the formative spirit,
Tests the forces of matter, hatreds and loves of the magnet,
Follows the sound through the air, follows through ether the
    ray,
Seeks the familiar law 'mid the grim wonders of chance,
Seeks the immutable pole in the phenomena's flight.

B. N. John.

The Restless Universe

———

CHAPTER I

The Air and its Relatives

IT is odd to think that there is a word for something which, strictly speaking, does not exist, namely, " rest ".

We distinguish between living and dead matter; between moving bodies and bodies at rest. This is a primitive point of view. What seems dead, a stone or the proverbial " door-nail ", say, is actually for ever in motion. We have merely become accustomed to judge by outward appearances; by the deceptive impressions we get through our senses.

We shall have to learn to describe things in new and more accurate ways. We say, " The air in the cinema is ' bad '; the hill air is ' good '." But goodness or badness is not a property of the *air*, but of what is mixed with it—dust, soot, water-vapour, &c. Now, just what *is* air?

1. *Air Pressure.*

We are using air all the time as we breathe in and out, and, as we know, our lungs get from it the oxygen we need to keep us alive. This is Chemistry, which we shall leave till later.

By keeping our mouths shut we can puff out our cheeks with the air. This is Physics; in fact, it is the simplest possible experiment on air pressure. We can learn a good deal more about air, however, by using an air pump, say a bicycle pump. When we push down the top of the pump (the piston) we squeeze the air inside the pump together, i.e. we compress it. The air then pushes open a valve and finds its way into the tyre. You may have noticed that by the time you have finished pumping up, the pump is quite hot. You may think that this is due to friction, like the warmth you feel if you rub the palms of your hands together. True, some of the heat *is* due to friction, but most of it is due to something else. Here is an experiment to prove it. Unscrew the pump and just pump into the air. Now the pump scarcely heats up at all. In pumping up the tyre, therefore, we have an example of *heating by compression.*

A bicycle pump is a compression pump. There are also suction pumps, which suck the air out of a closed vessel. They are used, for example, in making electric lamps; the bulbs are pumped free of air to prevent the fine metal filaments burning away. As a result of modern advances in science, we can now make very efficient pumps indeed, which can remove

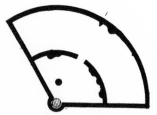

almost every trace of air (or gas) from a closed vessel. But we cannot go into the details of pump construction here; what we are interested in is the question, what *is* this substance air, which resists our attempts to squeeze it together and immediately spreads itself all through any space offered to it?

Think of a class at school just before the holidays. When the bell rings and the classroom door is opened, the children hurry out and in a minute or two the classroom is "rarefied" (almost empty). For a few minutes there is a low "concentration" of children in the corridors, but soon all of them find their way out at the main door. The holidays have begun, and inside the school there is a "vacuum" (1).

(1)

This is very like the idea which physicists have had of the nature of air, at least for the last hundred years or more. The physicist takes it for granted that air merely consists of a lot of small particles which he calls *molecules*. These molecules are flying about all the time and are continually colliding with one another and with the wall of the vessel containing them. If a part of the wall is movable, like the piston of an air pump (2), the continual hailstorm of particles will push it out, unless something on the other side of it prevents it moving out. The swarm of molecules tends to spread evenly throughout any space available, and if we try to make the air "thicker" (denser) by squeezing it together (as in the bicycle pump) the molecules will find their way through any chink or crevice until they are once more spread evenly (though more thinly) throughout the space.

(2)

You may say, " Yes, this is likely enough; but how are you going to prove it? Why do the molecules have ' that holiday feeling ' which makes them rush out at the door like children from school?"

Good. What we have done is to make a physical *theory*, and we must be prepared to produce evidence in favour of our theory.

The theory is called the *kinetic theory of gases*. The Greek word for motion is *kinema*, so that the name means that the theory is based on the motion of the molecules, i.e. on their restlessness. And the theory does not apply to air only, but is generally believed to hold for all its relatives, the other gases, such as oxygen and nitrogen, the constituents of air; hydrogen, which burns, and carbon dioxide, which puts out a flame; the poisonous carbon monoxide, the strong-smelling ammonia, the greenish-yellow chlorine, the so-called inert gases neon and argon, which do not react chemically at all, and others too numerous to mention.

At one time scientists were accustomed to draw a distinction between " permanent " gases, which could not be liquefied, and other gases which were known to be the " vapours " of liquid or solid substances, bearing the same relation to them as steam does to water. As lower and lower temperatures (3) were reached, however, one gas after another was liquefied; first carbon dioxide (at −78·5° C.), then air (at −193° C.), and other familiar gases. The last gas to resist liquefaction was helium; but at last Kamerlingh-Onnes succeeded in liquefying this gas also, at the extremely low temperature of −269° C.

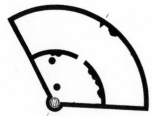

(E 969)

It is clear, then, that there is no real difference
between a gas and a vapour.   Conversely, every
liquid or solid substance can be vaporized, i.e.
transformed into a genuine gas, by the application
of high temperatures.   There is no substance, not
even iron or gold or platinum, that will not melt
and vaporize if the heat is great enough.   One of
the most difficult metals to vaporize is tungsten,
the boiling-point of which is estimated to be 4800° C.

Thus the concept of a gas includes every substance whatever—provided the temperature is sufficiently high; and the kinetic theory of gases is
believed to apply to all substances in the gaseous
state.

We must now bring forward evidence in favour
of the kinetic theory.

2. *Collisions and their Effects.*

Can we suggest any other theory?   For example,
we might think of air as consisting not of particles
rushing about all the time, but of particles which
have come to rest in the vessel, with forces of repulsion acting between them.   If the volume were increased, the particles would be free to expand under
these mutual repulsions.

A theory, to be of any real use to us, must satisfy
two tests.   In the first place, it must not make use
of any ideas which are not confirmed by experiment.   Special assumptions must not be dragged in
merely to meet some particular difficulty.   In the
second place, the theory must not only explain all
the facts we know already, but must also enable us

(3)

to foresee other facts which were not known before and can be tested by further experiment.

Now we shall consider our alternative theory. The assumption that the particles of air repel one another does not even agree with all the facts we know already, and has quite definitely been dragged in here to meet a particular difficulty. For it is well known that air can be liquefied by cooling and compression. That is, when the molecules of air are brought very close to one another, they *attract* one another and stick together. But two drops of liquid air should strongly repel one another, if each particle of the one repelled each particle of the other originally, that is, when the particles were farther apart. The repulsion theory will not do.

The kinetic theory, on the other hand, depends on something that we already know to be true, namely, the laws of the mechanics of moving bodies, in particular, the *law of inertia* and the *laws of collisions*.

Most people associate the word *mechanics* with workshops and machinery. Here, however, we are not concerned with the mechanics of lathes and milling-machines; but with a branch of science which originated from astronomy and which deals with moving bodies and the laws which govern their motion. Unfortunately, bodies on the earth are so crowded together and are subject to so many uncontrollable influences that it is difficult to observe the laws of motion in all their purity and simplicity. On the contrary, it is necessary to devise very artificial experiments, which the reader may vaguely remember from tedious physics lessons—experiments with pendulums (4) or apparatus for recording the fall of bodies, and so on. Anyone who thinks he has already heard enough about the laws of motion can omit the rest of this section—perhaps he will come back to it later.

For the benefit of other readers, however, we shall give a brief account of the most important laws of mechanics. These laws have been known since the time of Galileo, who was the first to state the ideas of velocity, acceleration, mass, force, &c., clearly and to illustrate their meaning by examples. As our example we shall take the familiar game of billiards (5).

(4)

The field where mechanics has really demonstrated its value is the theory of the motions of the heavenly bodies founded by Newton. Having stood the test in the heavens, it has, so to speak, been brought down to earth again to explain terrestrial phenomena.

The first fundamental law of mechanics is the *law of inertia*, which states that every body which is free to move without interference from other bodies retains the motion which it already has.

Clearly a statement like this is difficult to test by experiments on the earth. For how are we ever to isolate a body so that it is free from all outside influences? At best we cannot get rid of gravity, the attracting force of the earth itself. In the case of the billiard ball, however, the condition is at least partly fulfilled. As gravity acts vertically downwards, it can have no effect on the horizontal motion

(5)

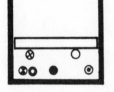

of the ball; and, apart from the action of the cue, no other forces are involved except a little friction and air resistance. The game of billiards, in fact, is just a succession of applications of the law of inertia. The stroke gives the ball a certain velocity, and it goes on rolling with this velocity long after the completion of the stroke. The same thing is exemplified in any ball game, and in many other cases; for example, a motor-car does not stop the instant the engine is switched off.

Inertia, however, is a property which can exist in different amounts. Suppose we replace the ivory billiard ball by a light ping-pong ball. A stroke which would make the billiard ball run slowly for a long distance over the green cloth gives the ping-pong ball a great velocity which dies away very rapidly, so that the ping-pong ball may even come to rest in a shorter time than the billiard ball. The property on which this difference of behaviour depends we call the *mass* of the ball. Here we are using the word *mass* in an *artificial* sense, as a measure of inertia, not, as in ordinary speech, as a quantity of substance. The *inertia* of the billiard ball is greater than that of the ping-pong ball; the heavier ball is less accelerated than the lighter ball when both are started from rest by equal strokes, but, on the other hand, the heavier ball retains its velocity more persistently against the frictional forces than the lighter ball does.

The reader must understand quite clearly that "heavy" does not mean the same as "inert". The weight of a body acts vertically downwards. If we lay the two balls in succession on the pan of a kitchen spring-balance, they compress the spring by different amounts, owing to their different weights (6). In the stroke with the billiard cue, however, gravity does not come in, as it has no effect in a horizontal direction. Here the differing results of equal strokes arise from the difference as regards persistence of motion or inertia. *Weight* is a measure of gravitational force, *mass* a measure of inertia.

(6)

To determine the mass of a body we might expect to have to make some such experiment as the following. We set up an apparatus which enables us to give each ball an equal impetus, say a pendulum with a hammer-shaped bob which is always made to fall from the same height before hitting the ball (7). If we use balls of all possible kinds, either solid or hollow, made of lead, brass, wood, celluloid, &c., but all of the same size and all equally smooth, we find that the velocities acquired by them are all different. We say that a ball which is driven off at only half the velocity of another has a mass twice that of the latter, and so on.

Fortunately, however, such experiments (which, it is safe to say, could not be performed very accurately) are not necessary. For there is a fundamental theorem, due to Newton, which states that weight and mass are always exactly proportional to one another. Both weight and mass, then, must depend on the same inner property of the substance.

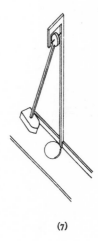

(7)

That this theorem is true is proved by the fact that in a vacuum all bodies fall at the same rate. True, a heavier body is attracted more strongly by

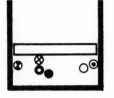

the earth, but its greater inertia enables it to resist acceleration to an exactly corresponding extent. The fact that the two effects just balance one another can be tested very accurately by means of experiments with pendulums or similar apparatus (8). Here again the driving force is the weight, and the inertia always acts against it. It is found that pendulums of the same length with bobs of differing weight perform exactly the same number of oscillations in a given time.

The deeper meaning of this equality of weight and mass did not merely escape the notice of the great Newton, but was unsuspected during the next two centuries, and has only in our own time been revealed by Einstein's theory of gravitation. This, however, lies outside the scope of the present book.

We shall always use the words weight and mass interchangeably. As regards units and numerical values, to be sure, a distinction must be made. The gramme (gm.) is the (scientific) unit of mass, and was chosen in quite an arbitrary way for practical reasons. The weight of this mass of 1 gm. is the force with which it is pulled down (accelerated) by gravity. As a body falling freely is found to acquire an acceleration of 981 cm. per sec. per sec., the weight of 1 gm. is 981 force units or dynes. That is, a dyne is the weight of 1/981 gm. or about 1/1000 gm. (1 milligram (mgm.)).

The product of the mass of a body and its velocity we call the *momentum* of the body. If one billiard ball collides with a second ball, part of the momentum of the first ball is transferred to the second: the momentum is distributed in some way or other between the two balls, but its total amount remains the same. This is just another way of saying that the centre of gravity of the system of two balls continues to move on unchanged in a straight line both before and after the collision. We illustrate the collision of the balls by a drawing of a piece of film showing a number of successive positions of the balls and of their common centre of gravity (9).

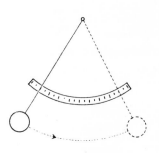

(8)

Forces like gravity, which act continuously all the time, may be imagined as made up of a great number of very small impulses. Each impulse alters the momentum by an imperceptible amount, but in time the effect becomes appreciable, and we have the law of motion

Time-rate of Change of Momentum = Force.

This is Newton's original statement of the law; we shall use it later, because it is readily adapted to the modern modifications of mechanics which have arisen from the theory of relativity (Chapter II, p. 73). As a rule, however, the law is stated in another way. As momentum is the product of mass and velocity, and the time-rate of change of velocity is acceleration, we may write

Mass × Acceleration = Force.

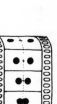
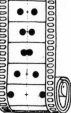

(9)

Here we have taken it for granted that mass is a constant. As we shall see later (p. 72), however, this assumption proves false in the case of rapidly moving bodies.

If we know how the force depends on the position

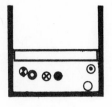

of the body under consideration relative to other bodies acting on it, we can calculate what the motion of the body will be. This is the root idea of Newton's mechanics, by means of which he was able to calculate the orbits of the heavenly bodies. The success with which the astronomer can predict the positions of the planets is the strongest proof of the truth of the laws of mechanics. In applying these laws to the motions of gas molecules we stand on safe ground. Before we do this, however, we must discuss one other concept, that of *energy*.

In everyday life this word is used in a great many ways, e.g. we may use it as meaning a person's capacity of forming decisions and acting on them. In science the word is used in an artificial sense. Energy means capacity for doing work, expressed by a number.

If we lift a weight of 1 lb. a height of 1 ft., we do an amount of work which is called 1 foot-pound. If the lifting is done by a steam crane, a definite quantity of heat is used up in the boiler. This heat, then, is also a store of work; it is capable of doing a definite amount of work and is called a quantity of energy. At the electrical power station the heat energy is first transformed into an electric current; this brings electrical energy into our houses, which we pay for, just as we do for any other goods. For energy is a kind of commodity; it is indestructible, although it may change its form.

When climbing a hill in a motor-car we make the engine run quickly beforehand, so that the car rushes forward and its momentum helps to carry it

up the hill. Even if the engine were throttled down at the beginning of the ascent, the car would climb a certain way by its momentum alone. A motion can therefore do work in raising a weight; thus it is a form of energy, and we speak of energy of motion or *kinetic energy*.

The kinetic energy of course depends on the velocity and the mass of the body, but it is not the same as the momentum. We can test this by finding what starting velocities must be given to different bodies if they are to do the same amount of work. If, for example, we shoot at a target in which the bullets remain embedded, the kinetic energy of the bullet is transformed into heat, and can thus be compared indirectly with the work done in raising a weight. We find that doubling the velocity of the bullet has just the same effect as quadrupling its weight: in general, the heat energy developed is proportional to the mass and to the square of the velocity. Half the product of the mass (m) and the square of the velocity (v) is called the " kinetic energy " (E): that is,

$$E = \tfrac{1}{2}mv^2.$$

Physicists formerly considered this to be the simplest and most elementary form of energy, and attempts were made to reduce all other forms of energy to it by regarding them as the kinetic energies of hidden motions.

In the case of heat this met with complete success, the result being the kinetic theory of gases, which we must now discuss. In the case of the other forms

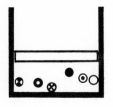

of energy, particularly electricity and magnetism, such attempts persistently made have met with no great success. To-day the tendency is to proceed in the opposite direction and regard the electromagnetic form of energy as the fundamental one and the energy of motion as derived from it. We shall come back to this later.

3. *The Motions of the Molecules.*

If we had a billiard table with perfectly elastic cushions and imagined the ideal case in which no extraneous retarding forces, such as friction or air resistance, were present, a ball once struck would fly on for ever in a zigzag path. If a number of balls were set in motion, each one of them would fly in a similar path, but occasionally two balls would collide. If the balls were perfectly elastic, there would then be no loss of momentum or energy on the whole; the balls would rebound in different directions, but their dance would go on unceasingly.

This is the picture which the kinetic theory of gases gives us, only in three dimensions instead of two. The molecules are regarded as perfectly elastic balls, which rebound without loss of energy from one another and from the walls of the containing vessel. Once the molecules are set in motion in any way, their inertia prevents this motion ever coming to an end.

The molecules, however, must not be thought of as a " perpetual-motion " machine, producing any amount of energy out of nothing; on the contrary, they really form a storehouse of energy. Their tendency to fill the whole of any space offered to them is an obvious indication of this stored-up energy. If there is a hole in the wall of the containing vessel, the molecules of course find their way through it. Again, the pressure which the gas exerts on the wall is just the sum of the small blows rained on it continually by the particles.

We shall now actually see the dance of the molecules in our first " film ". Of course this Film I is immensely magnified. Real molecules are far too small for us to see, even if there were not far too many of them to follow. But everyone is quite used to seeing countries and continents depicted on a very small scale in maps, and here we are just using the opposite device of representing very small objects on a greatly magnified scale.

First we see the molecules flying about and colliding with the walls of the vessel. Then a hand appears and pushes down the piston. Now the molecules have less room to fly about in, and each of them collides more often with the piston, so that the pressure required to reduce the volume further becomes greater and greater.

At the same time we see that the molecules all move more and more quickly. Why? Because each time they hit the *moving* piston they get extra momentum and energy (from the hand pushing the piston down), and bounce off faster than they came. This might be thought of no consequence, for the piston could be pressed down very, very slowly. True, the energy transferred to the molecules at each collision would then be less, but, on the other

hand, more time would be taken to reduce the volume to the same extent. Hence as the effectiveness of each collision falls off the number of collisions rises, in the same ratio. That is, the average increase in the speed of the molecules depends on the reduction of volume only and not on the speed of the piston (the rate at which the volume is reduced).

In return for the work done in compressing the air, we get an increase in the velocity of the molecules, and the real significance of these molecular motions is now apparent. They mean *heat*, for, as we know, a compression pump becomes hot in working.

The heating, not only of air or any other gas, but of any liquid or solid, just means that large-scale mechanical motions are transformed into the invisible motions of the molecules. No special significance attaches to the word " invisible ". The point of real importance is that the heat motions of the molecules occur " at random ", to such an extent that these motions cannot altogether be reduced to useful ordered motions. For example, the efficiency of a steam-engine rarely reaches the miserable figure of 30 per cent. The molecules are not like a gang of workmen who understand orders and work together according to plan (10); they are like a flock of sheep which the shepherd has some trouble in controlling even with the help of his dog (11). The engineer, who has cunningly contrived to make the blind and deaf molecules in their mad, senseless rush drive an engine, may well feel proud of himself.

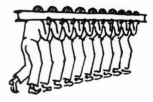

(10)

The restlessness of the very small parts of the universe, then, is a matter of very practical concern. Or rather, it shows us clearly what is practical and what is not. For it is here that human ingenuity breaks down. No matter how large man builds his machines, he can never surpass the degree of efficiency permitted by the haphazard nature of molecular motions. The best he can do is to account exactly for the energy lost. In the modern theory of heat this is done by statistical methods, which apply whenever we have to deal with a large number of random occurrences. These we shall now discuss.

(11)

4. *The Laws of Chance.*

How is it that chance plays a part in so exact a science as Physics? If we admit the influence of chance, we are surely denying the strict accuracy of the laws of nature?

To resolve this contradiction we must consider a little more definitely what we mean by a law of nature. We need only mention a simple mechanical phenomenon, say the firing of a shell from a gun (12). The path of the shell is determined by certain laws. As every schoolboy knows, if the resistance of the air did not exist the path would be a parabola. Can we say, then, where the shell will fall? No, we must also know the direction in which the gun-barrel is pointing, and the velocity of the shell as it leaves the gun.

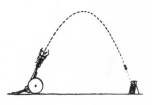

(12)

These " initial conditions ", as we may call them, clearly have nothing to do with the " laws of nature ". But if we are going to *apply* the laws of nature we

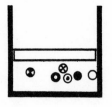

must know the initial conditions as well. Otherwise we are not in a position to make any useful predictions.

In our heat problem, where the same processes occur over and over again, we have the remarkable compensation that chance is itself subject to law. This may seem paradoxical, but there is no doubt whatever of its truth. If it were not so, the uncommon achievement of " the man who broke the bank at Monte Carlo " would never have been commemorated in song. Many philosophers have racked their brains over the deeper meaning of the laws of chance, but they have left gamblers, insurance companies, physicists, and others to apply these laws and satisfy themselves that they lead to correct conclusions.

The physicist, then, accepts the laws of chance as laws of nature, although he cannot say what their ultimate metaphysical meaning may be. The great French mathematician, Henri Poincaré, has stated the position with typically French penetration, putting the words into the mouth of an eminent physicist:

" ' Tout le monde y croit fermement parce que les mathématiciens s'imaginent que c'est un fait d'observation et les observateurs que c'est un théorème de mathématiques.' Il en a été longtemps ainsi pour le principe de la conservation de l'énergie."*

But, honestly, do we really know the ultimate meaning of other natural laws which we do regard

* Preface to *Thermodynamique*, Paris, 1892.

as exact, such as the law of inertia or Newton's celebrated law of gravitation?

As a matter of fact, the most recent development in physics, quantum mechanics (which we shall discuss later), has shown that we must drop the idea of strict laws, and that *all laws of nature are really laws of chance*, in disguise.

Statistics are rather suspect in everyday life; the word calls up ideas ranging from the gambler and his unpleasant end to the most respectable insurance company. The malicious even say, " You can prove anything by statistics ". Statistics are perhaps not always applied in legitimate ways. Some kinds of business, however, depend directly on the reliability of statistics, e.g. life insurance. With the help of the theory of probability the actuary calculates the premiums for the various ages from tables of mortality. A mistake on his part might ruin the insurance company. People who insure their lives thereby demonstrate their faith in the correctness of the mortality tables and the reliability of the actuary's calculations.

The physicist who seeks to apply statistical theory to *his* problems is at least equally sure of his ground. His raw material consists not of empirical tables, but of simple assumptions about ." equal probabilities ". We all know what this means from our experience of dice-throwing. Each of the six faces of the die has an equal chance of coming on top. Otherwise the die is false; that is, its centre of gravity does not lie exactly at the centre, or else it deviates in some other way from the ideal form.

If, after careful examination, no such fault is found, we may expect that if a great many throws are made, one particular throw, say ⟦::⟧, will occur in one-sixth of the throws. This has been confirmed by actual experiments with dice.

The theory of probability tells us what we may expect to happen in complicated cases, e.g. the chance of making a particular throw when several dice are thrown together, or the chance of a deviation from the mean occurring (say the chance of throwing ⟦::⟧ not 100 times in 600, but only 90 times). This "science of ignorance", like any other method of investigation, is justified by its results. The greater the number of cases considered, the more accurate the predictions we can make. Thus the physicist is in a very favourable position, for in physical phenomena the number of particles is generally enormous.

The simplest physical case is that which we are particularly concerned with here, namely, a gas. Here we have an immense number of similar molecules, and we neither know, nor care very much, what happens to any individual molecule, but we do want to know all we can about the *average* properties of the molecules—for example, their mean density, mean velocity, and so on.

Any one picture of Film I is, so to speak, a snapshot of the molecules, and may be thought of as giving an "initial position". The particular appearance of the picture is thus quite "accidental", but that does not mean "lawless". For example, everyone will readily admit that in a series of such snapshots only very few will show almost all the

molecules in the right-hand half of the containing vessel. In an overwhelming majority of cases the molecules will be fairly evenly divided between the right-hand half of the vessel and the left-hand half.

Now why is this? Let us think of only very few molecules, say four, and call them John, Edward, William, and George. In how many ways can they arrange themselves so that two of them are in the right-hand half of the container and two in the left-hand half? Clearly there are six ways, namely:

Left.	*Right.*
John and Edward.	William and George.
John and William.	Edward and George.
John and George.	Edward and William.
Edward and William.	John and George.
Edward and George.	John and William.
William and George.	John and Edward.

On the other hand, there is only *one* way of arranging themselves so that all four of them are in the right-hand half of the vessel and none in the left, and only *four* ways such that three are on the right and one on the left (for this one can be any one of the four).

Even when there are only four molecules, the *uniform* distribution is distinctly the commonest arrangement, and with more molecules it becomes much more definitely the most abundant one.

The problem of calculating the relative probabilities of the different arrangements is not very difficult and can be reduced to other well-known problems, for instance, that of tossing pennies.

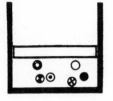

Collect together ten pennies and throw them up. Each of them will come down heads or tails. Count up how many come down heads. Suppose it is three, so that three are heads and seven tails. Throw the whole set of coins up a large number of times, say 500, and count up how many times out of the 500 you get no heads and 10 tails, 1 head and 9 tails, 2 heads and 8 tails, 3 heads and 7 tails, and so on to 10 heads and no tails. Before you do the experiment, I will tell you the result. The numbers you will get if you do it 500 times are very near to these figures:

Heads: 0	1	2	3	4	5	6	7	8	9	10
Tails: 10	9	8	7	6	5	4	3	2	1	0
0 to 1	5	22	58	103	123	103	58	22	5	0 to 1

It will be very surprising indeed if you get results differing from any of these figures by more than 8, .say, at the outside.

Now let " head " stand for a molecule in the right-hand half of the vessel, and " tail " for a molecule in the left-hand half. Then the little table at once gives the frequency of the occurrence of the different molecular distributions in a set of 500 snapshots of the ten molecules. We see that the distribution 5 heads, 5 tails—i.e. equal numbers of molecules in the two parts of the vessel—has the highest probability (123 out of 500) and that the frequency of the distribution diminishes as the distribution becomes more and more uneven.

The preponderance of the uniform distribution becomes very much more marked if the number of molecules is greater. For example, if there are 100 molecules, the number of ways we can place them so that 50 are on the right and 50 on the left is enormously greater than the number of ways we can place them with 90 on the right and 10 on the left; in fact it is some 10,000,000,000,000,000 times as great.

We are now on the track of a statistical law. We are pretty safe to bet that in any particular case the distribution will be nearly uniform, for this is so much more likely than any other arrangement. This is all the truer for a gas, owing to the immense number of molecules. Now we can understand why no one has ever been suffocated just because all the molecules of air happened to be in the other half of the room at once. It must be conceded that a tragedy of this kind is *possible*. But it is exceedingly rare—so rare that the whole history of the human race, from the " missing link " to our ultimate posterity, would not give sufficient time for it to happen even once.

Yet it is just these rare occurrences, these deviations from the mean, that are of particular interest, as they prove the truth of statistical ideas; it is these that the physicist seeks. We shall discuss them later (p. 36). Meanwhile we shall confine ourselves to " normal " occurrences.

5. Temperature and Molecular Velocity.

As with many scientific notions, the idea of temperature arose directly from our sensations. We can feel that one body is hotter than another. Then we say it has a higher temperature. This certainly

does not mean that it contains more heat.   We may burn our tongue by taking a spoonful of hot tea, but can easily drink many cups which have cooled down a little.   The heat content of the spoonful of tea is much less than that of the several cups of cooler tea.   Temperature is therefore a quality;   heat a quantity.   To detect " temperature ", a thermometer is an improvement on our fingers.   The simplest case is that of equality of temperature between two bodies;   it is clear that this does not depend on the scale of the thermometer used.   We take different gases, for instance, and bring them all to the same temperature, using any sort of thermometer we like.   Then they are " in thermal equilibrium ", that is, they do not exchange heat. What has happened to the molecules when this " thermal equilibrium " exists?

A simple experiment leads us to the correct answer.   If equal volumes of two different gases are at the same temperature, they have always the same pressure.   Now the pressure is due to the rain of particles on the walls, and it is therefore proportional to the number of particles and to the square of their mean velocity.   For if the velocity is doubled, not only is the effect of each collision doubled, but the number of collisions in a given time is also doubled, so that the pressure is quadrupled.

As we have already seen (p. 13), half the product of the mass of a particle and the square of its velocity is a measure of the energy of motion of the particle and is called its *kinetic energy*.   Using this expression, we can say that the pressure of a gas is proportional

to the mean kinetic energy of its particles. Thus we are led to the idea that equality of temperature of two gases means nothing else than equality of the mean kinetic energies of the molecules. This can be confirmed by a detailed investigation, which shows that two gases brought into contact or actually mixed together exchange energy until the mean kinetic energy of one kind of molecule is equal to the mean kinetic energy of the other kind.

The natural definition of temperature, which has been adopted in science, is the mean kinetic energy of the molecules of a gas. But as this would be difficult to determine experimentally we prefer to measure the temperature by the pressure of the gas. The scale is usually chosen so that the interval between the freezing-point and the boiling-point of water is 100 degrees. But what number are we to give to the lower mark, the freezing point? We have agreed that the number representing the temperature is also to measure the mean kinetic energy of the molecules. Obviously there is an *absolute zero* of temperature, corresponding to the state where all the molecules are at rest and the pressure has dropped down to nothing. By observing the rate of decrease of pressure on cooling, it is found that the absolute zero lies 273° below the freezing-point, so that the latter point is 273°. We call this scale the *absolute scale*, and speak of so many degrees absolute, or degrees Kelvin (°K), since we owe this scale to Lord Kelvin. Ordinary room temperature is about 290° K. What, then, is the mean velocity of the molecules of a gas at room temperature? This cal-

culation involves only some knowledge of mechanics and of the theory of probability. We shall spare the reader these calculations; no doubt he will be prepared to trust the mathematicians to do their sums correctly. It turns out that the molecules of the air in the room are actually rushing about at a mean speed equal to that of a rifle bullet—about 1300 ft. per second!

When the kinetic theory of gases was first suggested over a century ago, this result excited much surprise. It might, for example, be objected that we ought to smell gas instantly anywhere in the house if the cook happens to leave the tap of the gas stove on. This, unfortunately, is not what happens, otherwise many tragedies might be prevented. Well, then, does not this slow spreading of the smell of gas contradict the statement that the molecules have very high speeds? There are other slow " creeping " motions in gases. These are all related to one another, and will now be explained.

6. *The Conduction of Heat.*

Have you ever thought why you cover yourself with blankets at night?—Because we want to keep warm, of course!—But the blankets do not themselves give out heat?—Of course not, that is not necessary; we are ourselves a kind of portable stove. —Then the blankets are to prevent you losing heat, then? Are they poor conductors of heat?—No, they conduct heat better than air does; the physicists have proved it!—Why do you not cover yourselves with air, then? That would be easier and cheaper

too!—Yes, that would be all very well if the air would keep still. In reality, however, warm air tends to rise, dragging cold air after it. That is why we cannot keep warm without blankets. The blankets slow down the current of air, and along with the air caught between the fibres form a sort of stationary jelly—in this sense we do cover ourselves with a blanket of air, and so utilize the fact that the air is a bad conductor of heat.

But how does this come about? If the molecules· are really rushing about with a speed of 1300 ft. per second, how is it that any local heating, that is, any increase in mean velocity at a particular place, is not propagated at the same high speed?

The answer, as has no doubt occurred to the reader, is that the molecules get in one another's way. We see this happening sometimes in Film I. Though a molecule has a high speed, it cannot go far without hitting another one and being deviated from its original direction, just like a billiard ball. Clearly this prevents the rapid spreading of any peculiar condition like high velocity (i.e. high temperature), and the smaller the average length of the free path between two successive collisions (" mean free path "), the more slowly will the spreading take place.

The mean free path depends on two things, namely, the number of molecules per cubic centimetre, and their size. The greater the number of molecules, and the greater the size of each, the shorter is the mean free path.

This result is important. For we see that it is possible to get information about the number and

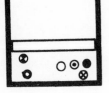

size of the molecules from the rate of propagation
of a difference of temperature (i.e. from conductivity
measurements) or from the rate of spreading of, say,
coal gas through air (so-called diffusion). The
physicists have actually done this. But a lot of
mathematics is needed, and after all the results
are not very exact.

In science we do not always advance from the
simple to the complicated. Often the reverse: a
result is obtained in an extremely indirect and
troublesome way, and later some simple and direct
way of demonstrating it is found.

The present-day physicist has available simple
direct experiments for finding the velocities of
molecules as well as their size and number.

7. *Molecular Beams.*

The physicist uses a method characteristic of the
whole of modern physics, namely, the production of
molecular beams.

We shall see later that most of what we know
about the smallest particles depends on setting them
in rapid motion—making a " beam " of them and
causing it to hit some other matter, and observing
what happens.

The practical instruments for producing molecular beams are quick-working high-vacuum pumps.
(These pumps are themselves based on phenomena
discovered by the help of the kinetic theory of gases,
the most important being diffusion.) In a very high
vacuum, the molecules fly from wall to wall of the
containing vessel without colliding with one another.

If a vessel containing gas at an appreciable pressure is connected with a high vacuum by a tube, the molecules will rush through the tube and emerge as a well-defined beam. Then to prevent them spoiling the vacuum, all we have to do is to make them stick to the opposite wall—at the same time demonstrating the existence of the beam. The " fly-paper " we use is simply low temperature. The wall of the vessel which is to act as a molecule-trap is kept cool, and it gradually collects a spot of the condensate of the molecules which arrived in the form of a beam.

The fact that this experiment works is, of course, in itself a welcome confirmation of the kinetic theory of gases; it shows directly that the tendency of the gas to expand is really just the tendency of the molecules to keep on flying in straight lines according to the law of inertia.

Now we can actually measure the velocity of molecules. One method, shown in Film II, is due to Stern, and consists in rotating the whole molecular-beam apparatus rapidly. In the centre is the source of molecules, an electrically-heated platinum wire on which a layer of silver has been deposited. The silver particles evaporate and are shot off in straight lines in all directions. Most of them are caught on a (circular) screen placed in the centre of the vessel. Some, however, fly through an opening in the screen and form a molecular beam. When the apparatus is at rest the molecular beam gives

(13)

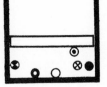

rise to a spot of silver on the outer wall, just opposite the opening. This spot appears in all the pictures of Film II. The series of pictures begins at the instant when the apparatus begins to rotate. The particles take no notice of the rotation, of course, but go on flying in straight lines, and accordingly fall on a part of the receiver farther to the left (relative to the direction of rotation), where a second spot is formed. From the distance between the two spots and the velocity of the rotating receiver we can at once deduce the velocity of the molecules. The result agrees exactly with that calculated from the kinetic theory of gases.

Other more refined mechanical methods have been used, e.g. one after the pattern of the well-known method for measuring the velocity of light. Two identical toothed wheels rotate on the same axis (13) (p. 29). If the wheels are at rest, a beam of molecules passing between two teeth on the first wheel will also pass between two teeth of the second wheel. If the wheels are made to rotate faster and faster, a molecule which just slipped through the first wheel will be caught by a tooth of the second wheel, but if the number of revolutions is still higher it will pass through the next gap. This method can be so refined that we do not merely get the mean velocity, but can establish the fact that there are molecules of every possible velocity present and ascertain how many there are of each velocity.

There will be a group of very fast molecules and other groups of slower velocity, down to zero. The law of distribution, which tells us how many molecules there are in each group, can be calculated theoretically, and experiment shows that this theoretical prediction is in good agreement with the facts.

Further, the apparatus can be used to produce a molecular beam of almost uniform velocity. We shall make use of this later (p. 154).

(14)

8. *The Size and Number of the Molecules.*

Suppose we had a basket full of peas (14) and wanted to know in a big hurry how many peas there were. Counting would of course be much too laborious.

A quicker way would be the following:

We estimate the volume of the basket as being equal to a cube with the edge $b = 10$ cm.; then its volume is

$$V = b^3 = 10^3 = 1000 \text{ c.c.}$$

If each pea were a little cube with the edge a, the volume of each would be a^3, and therefore that of n peas would be

$$na^3 = V = 1000 \text{ c.c.}$$

(15)

Now we spread out the peas on a table (15) so that they are just touching each other, and then measure the occupied area, $A = 1600$ sq. cm., say. As the surface of each little cube, representing a pea, is a^2, the total area is $na^2 = A = 1600$. Dividing the other equation by this we get

$$\frac{na^3}{na^2} = a = \frac{V}{A} = \frac{1000}{1600} = \tfrac{5}{8} \text{ cm.}$$

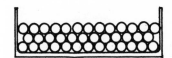

(16)

Of course the peas are more like spheres than cubes (16), but for the moment we can ignore this. We can safely assume that the calculated length $a = \frac{5}{8}$ cm. represents, roughly, the diameter of the pea. The total number of peas is then obtained,

from $na^3 = V = b^3$, as $n = \left(\dfrac{b}{a}\right)^3 = (10 \div \tfrac{5}{8})^3 = 16^3$,

which is about 4100. Figure (17) shows how the volume V is cut into 16 sheets, which when put together side by side give the area A.

This illustrates the main features of the methods used by physicists for counting and measuring molecules. Their method consists in first compressing a certain quantity of molecules together to form a solid body and measuring its volume, and then making the same molecules (or a known portion of them) form a coherent surface layer one molecule thick and measuring the area of this layer. The ratio volume/area then gives the thickness of the layer of molecules and hence an approximate value for the diameter of a molecule. Knowing this, we can find the volume of a molecule, and if we divide the total volume by this we get the number of molecules.

The first step is easily carried out experimentally, for all substances become solid on cooling, and it is quite reasonable to suppose that their structure resembles that of a heap of tightly packed peas.

But now we have to spread out a portion of this substance in such a way that the resulting layer is " unimolecular ", and measure its area. This is not so easy.

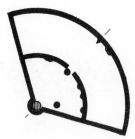

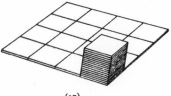

(17)

We may mention that it *is* actually possible
to produce films of oil on the surface of another
liquid which are only one molecule thick, and to
prove that this is so. But this involves the use of
complicated organic substances, whereas we are concerned with simple gases.

Here we again find the method of molecular beams
useful. We produce a beam of silver molecules in a
vessel where there is a high vacuum. Then the
vacuum is destroyed by letting in a small quantity
of the gas in which we are interested, say air. The
beam of silver molecules is now weakened by
collisions. If the air admitted is very rarefied, it will
act as if all the molecules were in one and the same
plane, for it will only very seldom happen that there
are two molecules both exactly in line with the
direction of the beam (18). A bullet fired from a gun
among straggling trees may hit one tree and pass
through it, but has very little chance of meeting
another tree. We now let in, say, enough air to cut
down the beam of silver molecules by half; if the
cross-section of the beam has the area a (say 2 sq.
mm.), then the area covered by the air molecules
(if they were all laid side by side) will be $\frac{1}{2}a$ (or
1 sq. mm.).

We have next to ascertain how large the granule
of solid air consisting of these molecules would be.
The pressure in our apparatus is some small fraction
(P) of the normal atmospheric pressure. If the
volume of the beam is $h \times a$ c.c., it will contain
$h \times a \times$ P times as many molecules as 1 c.c. of
ordinary air. Now it is known from special experi-

(18)

ments that 1 c.c. of ordinary air condenses to a solid granule of volume 1/2000 c.c. Therefore our column of air would give a granule of volume $\frac{1}{2000} \times h \times a \times P$. Dividing this by the area $\frac{1}{2}a$, we find, for the diameter of the molecule, the approximate value $d = \{\frac{1}{2000} \times h \times a \times P\} \div \frac{1}{2}a$; i.e. $d = \dfrac{hP}{1000}$ cm. If, for instance, $h = 5$ cm. and

$P = \dfrac{1}{200,000}$ of the atmospheric pressure, we find

that d is $\dfrac{5}{1000 \times 200,000}$, i.e. $\dfrac{1}{40,000,000}$ cm.

Now we could also calculate the number of molecules in a cubic centimetre of air at ordinary pressure. But we should have to write down so many noughts that the number would be quite unintelligible. We shall therefore introduce an abbreviated notation commonly used in physics. For 100 we write 10^2, for 1000, 10^3, and so on, that is, 10 with an upper figure (exponent) giving the total number of zeros. In the same way we write 1/10 as 10^{-1}, 1/100 or $1/10^2$ as 10^{-2}, 1/1000 or $1/10^3$ as 10^{-3}, and so on.

We may then state our result as follows: the diameter of an air molecule is about $1/(4 \times 10^7)$ or $\frac{1}{4} \times 10^{-7}$. As $\frac{1}{4} = 0.25 = \dfrac{2.5}{10} = 2.5 \times 10^{-1}$, we may also write the number in the form 2.5×10^{-8}.

The whole of molecular physics is dominated by this " order of magnitude " 10^{-8} cm., or one hundred-millionth of a centimetre. This magnitude has been given a name; we call it an Ångström

unit (Å.), after a Swedish scientist. A molecule of air, then, is about $2 \cdot 5$ Å. across.

We now easily obtain the number (n) of molecules in 1 c.c. of ordinary air. For as the volume of one molecule is d^3, these n molecules packed together form a solid of volume nd^3, which, as we know, is equal to $\frac{1}{2000}$ c.c. Therefore n is $\frac{1}{2000\,d^3}$ and, introducing the value of d, namely, $\frac{1}{4} \times 10^{-7}$ cm., we find that

$$n = \frac{1}{2000 \times \frac{1}{64} \times 10^{-21}} = \frac{64}{2 \times 10^{-18}} = 32 \times 10^{18}.$$

This fundamental number, which can also be written in the form 3×10^{19}, was first determined by Loschmidt, and accordingly is often called *Loschmidt's number*.

It is an exceedingly large number, 3 with no less than 19 noughts following it! However, there is no sense in trembling with awe before it; rather we should seek to gain a clear idea of how such an enormous number can be determined. That is why we have discussed the method in such detail above, for it is typical of all cases of the kind, which in future we shall deal with much more briefly. The essential point of the method is the splitting up of the required very large or very small number into factors which are themselves of moderate value and hence accessible to experiment. In our case we used the number 2000, determining the contraction of a gas on solidification, and the number 200,000, giving the reduction of pressure in the molecular beam

apparatus. These are both numbers which we can easily grasp the meaning of. We can think, say, of peas in a row; if a pea is $\frac{1}{2}$ cm. across, a row of 2000 peas would be 10 metres long, and a row of 200,000 peas 1000 metres long.

On the other hand, it is much more difficult to grasp the meaning of the number 4×10^7, which gives the number of molecules which when arranged in a row would come to 1 cm. If they were peas $\frac{1}{2}$ cm. across, say, instead of molecules, the row would be 200 kilometres long (125 miles), say about the distance from London to Birmingham! And a solid body of volume 1 c.c. would correspond to a cube full of peas with a side reaching from London to Birmingham!

9. *Pollen and Cigarette Smoke.*

Now are we really to believe all this, or is it just playing with figures?

Scepticism is wholly justified. A single phenomenon is no solid basis for a theory, nor is a single measurement a convincing determination of a value. It is essential to be able to make predictions and to confirm them experimentally in such a way that no doubt remains. This, as we shall show by means of a few examples, can be done in the present case.

The most speculative assumption we have made is that the laws of gases are not real " causal " laws, but depend on " chance ". Is there any direct proof of this?

In throwing dice or playing roulette or any other game of chance, we can be pretty sure that in the

long run the chances of all the possible events are equal. For example, anybody would be willing to bet high on the chance of the throw ⚅ coming up between 80 and 120 times in 600 throws. But nobody would be willing to bet so highly on the chance of the throw ⚁ turning up once and only once in six throws, although the probability of its turning up in a single throw is exactly 1/6.

We should accordingly expect to obtain noticeable deviations from the gas laws also, if only we could reduce the number of individual "events", for example the number of collisions involved.

Referring back to Film I, we see that in it only six molecules are drawn (it would have been too troublesome for the artist to show any more). In that case, however, the piston should not stay at rest, but should wobble up and down appreciably, as it is of course only supported by the impacts of the particles of gas. As regards this point our drawing is not strictly accurate. In reality the number of molecules is so great and their individual effects on the heavy piston so small that we do not see it shaking. But would it not be possible to use tiny little light pistons? Yes, this can actually be done.

More than a hundred years ago a botanist named Brown, on looking down his microscope, saw that the minute particles which detach themselves from pollen when the latter is thrown into water keep continually moving about like a swarm of bees. But it was only in 1906 that Einstein and Smoluchowski realized that this is a direct proof of the

reality of the molecular motions and of the truth of the statistical theory.

Of course the particles seen under the microscope are not actual molecules; they are still a thousand or ten thousand times as large. But they *are* what we just spoke of, namely, tiny pistons. It does not matter whether a liquid or a gas is used. If the blue smoke from a cigarette is examined through the microscope, thousands of particles are again observed. These float in the air and are struck on all sides by the air molecules; but as the particles are very small, the impacts of the molecules do not exactly balance one another; it often happens that there are more impacts, say, from the right than from the left, and the particle is made to quiver. It moves about in a wildly zigzag path, following purely statistical laws.

These laws are known; they can be used to calculate the number of air molecules in a cubic centimetre, e.g. by making a large number of observations of the time taken, on the average, by a particle in moving through a millimetre from right to left. The result is a brilliant confirmation of the number we previously obtained, namely, 3×10^{19} molecules per cubic centimetre.

Apart from exact measurements, however, no one who has ever seen the swarming points of light under the microscope will cast any doubt on the truth of the kinetic theory of gases.

10. *The Blue Sky and the Red Sunset.*

On clear winter days when all the dust and soot have been washed out of the air by snow the sky

shows its own true colour, a pure clear blue. On high mountains, above the vapours which rise from forests and human settlements, it is a still more beautiful darker blue. The higher we rise, the less air do we have overhead, and the darker is the blue of the sky. If we could get up above the atmosphere, the sky, even when the sun is shining, would without doubt appear as black as it does at night.

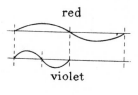

(19)

The sky is not the crystal sphere which the ancients imagined; the appearance of a blue vault is an illusion, and as there is nothing up there but air, the blue must be the colour of air. That is why we are interested in it. We want to know how a transparent substance like air can appear so brightly coloured.

Bodies which do not themselves emit light can only be seen when they reflect light which falls on them. This is true of air. When rays of sunlight reach the atmosphere, part of the light is reflected by the air and reaches our eyes (19). The question is, why is this light so highly coloured, and why is it blue rather than any other colour?

Later we shall discuss the nature of light in greater detail. Here we need merely recall that it is a wave motion, and that the different wave-lengths correspond to different colours. The longest light-waves (which have a wave-length of less than a thousandth of a millimetre) appear red to the eye. As the wave-length decreases we experience the colour sensations yellow, green, blue, and violet, in that order. The shortest violet waves which are still just visible have a wave-length about half that of the longest visible red waves (20).

red

violet

(20)

(E 969)

If waves of varying wave-length fall on the eye, we see mixed colours in all the variegated beauty with which Nature is adorned. Even the white of daylight is not a pure colour, but a mixture of waves of all wave-lengths. If this white light falls on a body, the body does not " look " white unless it reflects all the waves constituting white light uniformly. If, however, the body absorbs blue and green, say, it will appear orange. As a rule, therefore, colour in a body indicates that some light-waves are not absorbed by the body.

Usually this separation of absorbed light and reflected light takes place just at the surface of the body. This is why bodies can be ornamented by quite thin layers of paint. There are cases, however, where the separation takes place in the interior of the body, as in stained glass. Glass is in itself transparent to all the visible colours, and would therefore be " invisible " if feeble reflections from its surfaces or defects in its interior did not betray its presence. If metal in a finely divided state is added to the molten glass, the glass becomes coloured; for a separation of the light-waves takes place at the fine particles and the waves are more or less reflected.

In the case of very small particles whose diameter is of the same order of magnitude as the wave-lengths corresponding to visible light, we can see at once what the colour of the deviated light will be. For we have all had experience of wave motion when sailing. People who are subject to seasickness know quite well that they are most likely to suffer from this when the waves are about the same length as the boat. If the waves are much shorter they have but little effect; if they are much longer, they raise and lower the boat as a whole but do not make it roll or pitch. It is only when the bow is on a wave-crest and the stern in a wave-trough that a highly unpleasant rolling and pitching motion results (2 1). In this case, the motion of the boat will in its turn give rise to new waves, which travel out in circles round the boat.

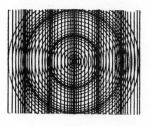

(21)

What we are interested in here is a fundamental phenomenon which the physicist calls " scattering " of the waves (22). Marked scattering occurs when the diameter of the body causing the scattering is about the same as the wave-length of the waves falling on it. This is true, not only for water waves, but also for any other kind of wave motion, whether of sound or light. Light of definite wave-length, then, is scattered most markedly by particles of a certain definite size. This is why glass containing very minute particles appears blue, while glass containing larger particles appears yellow, green, or red; for the small particles scatter the short-wave blue light more effectively than they do the light of longer wave-length.

(22)

We now return from the variously coloured glasses to the colour of the sky, which is really the colour of the air. Its explanation, however, involves still another idea, for, as we know, the blue of the sky is particularly fine when the air is especially free from floating dust particles and water vapour. How does this come about?

The explanation is provided by the kinetic theory of gases. We know that the molecules are never at rest, and are never distributed quite uniformly. There will always be small regions in the gas where there are more than the average number of molecules, and other regions where there are fewer molecules than the average. These regions have just the same effect on the light-waves as foreign bodies have. It often happens that a street on which the sun is brilliantly shining down suddenly looks as if it were covered with water. In reality it is only the effect of the reflection of the light at a heated layer of air. The slight difference in density is sufficient to cause the light to be reflected. In exactly the same way, the advance of light through a gas is always subject to slight disturbances, and secondary spherical waves are therefore produced wherever the density happens for the moment to be above or below the average.

If the region of momentary condensation corresponds in size to a particular wave-length, then light of this particular wave-length will be scattered to a greater extent than light of other wave-lengths. Now condensations of small extent will clearly be much commoner than condensations of greater extent.

This is best seen if we fix on a very small region which should on the average contain just one molecule (23). In reality, of course, it will sometimes contain no molecule at all, and sometimes two or more. Thus the density in these small regions will often vary by as much as 100 per cent. If, however, the region is so large that on the average it contains 10 molecules, it will often happen that the actual number of molecules present is 8, 9, 11, or 12. But this implies a variation of only 10 to 20 per cent in the density; a variation of density amounting to 100 per cent, i.e. a doubling of the number of molecules present, will be an extremely rare occurrence.

We now have the explanation of the blue colour of the sky. The short-wave blue light has a great many more chances of meeting with density variations of the right size than any of the longer waves have.

In the evening when the sun is setting the sky becomes pink, the clouds have their margins tinged with purple, and the sun sinks below the horizon as a blood-red disc. How does this colour transformation arise? It involves no new phenomenon, but merely arises from another aspect of scattering. If a sheet of glass, which when held in the hand looks red, is held so that we look through it at the sun, the sun appears greenish blue. For what we see then is the original, unscattered light, from which the red scattered light is missing; hence it has the colour " complementary " to red.

Similarly in the sunset sky: when the sun is low we can look at it with impunity, because the rays then have to pass through a long layer of air and are much weakened by scattering (24). It is only the red light, which is complementary to the blue scattered light, that reaches the eye and illuminates the edges of the clouds.

The blue of the sky and the red of the sunset are

(23)

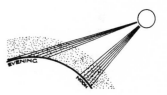

(24)

thus different aspects of the same physical process.

In physics, methods and apparatus for measuring the wave-length of light have been developed; we shall describe these later (p. 111). If such apparatus is pointed to the sky, we obtain data from which the number and size of the molecules can be deduced —in excellent agreement with our previous results.

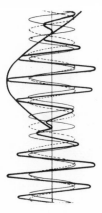

A number of other phenomena can be used for the same purpose. But they are incomparably less poetical in their associations than the blue of heaven and the crimson splendours of the sunset; they smell of nothing but the laboratory. And as they do not involve any fresh principle, we need not trouble ourselves about them.

11. *First Subdivision of the Molecules: Atoms.*

We have counted the molecules in the air—and we could do the same for other gases, e.g. for the familiar coal-gas which we use for cooking. We should obtain exactly the same number of molecules per cubic centimetre (at normal pressure and room temperature). Is this number, then, really invariable?

It is, so long as we are dealing with one definite gas, or else with a mixture of gases, such as is formed, say, if we let coal-gas mix with air. But if we put a lighted match to this mixture there is an explosion; a chemical change takes place, and this alters the number of molecules.

Instead of coal-gas, which is itself a mixture of gases, we prefer to consider hydrogen, and instead of air we shall take one of its constituents, oxygen. (These can be bought separately in steel bottles.)

When these gases are mingled they give an extremely explosive mixture. If some of the mixture is made to explode in a strong vessel, the chief residue is found to be water vapour, together with certain amounts of unused hydrogen or oxygen, according to which was present in excess. It is possible to adjust the quantities so that none of either gas is left, that is, only pure water vapour is produced.

Chemists have proved that in order to do this it is necessary to have the original gases present in perfectly definite proportions, namely, about 1 part of oxygen to 8 parts of hydrogen (by weight), these giving rise to 9 parts of water. We are not interested so much in this as in a certain physical fact. The resulting water vapour at 100° C., say, has by no means the same pressure as the original mixture of gases had at the same volume and temperature. The chemical combination has the effect of lowering the pressure by one-third.

This forces us to the conclusion that the number of water molecules is less by a third than the total number of molecules of hydrogen and oxygen originally present. For we know, of course, that the pressure is simply determined by the force and number of the impacts of the molecules on the walls of the containing vessel, or, more accurately, by the kinetic energy of the molecules and the number of them per cubic centimetre. The kinetic energy, however, is on the average the same if the temperature is the same; hence if the pressure is less by one-third the number of molecules must be less by one-third.

This is nothing to be surprised about. For the

hydrogen and oxygen have combined and formed water vapour. The number of happily married couples is only half that of the total number of brides and bridegrooms!—But stop, this agreeable picture will not do, for in that case the pressure should have fallen by a half, not by a third.

This third reveals that not only the molecules of compounds like water, but also those of gases which are elements, like hydrogen and oxygen, consist of *smaller particles.*

The simplest assumption is that each consists of two particles. These, as is well known, are called *atoms*; the atom of hydrogen is denoted by the letter H, that of oxygen by the letter O. For the corresponding molecules we then use the symbols H_2 and O_2. We can now very easily explain the fall of pressure by a third. The molecule of water consists of three atoms, either H_2O or O_2H. In the first case the explosion would take place according to the equation

$$2H_2 + O_2 = 2H_2O,$$

which means that two hydrogen molecules meet an oxygen molecule; they break up and form two water molecules. Or perhaps the reaction is the converse one,

$$2O_2 + H_2 = 2O_2H.$$

No, this is easily disproved. For we can find out what pressure the hydrogen and the oxygen, required to make water, exert separately (of course at the same temperature and in similar containers); we find that

the pressure of the hydrogen is twice that of the oxygen. Thus the explosive mixture of hydrogen and oxygen which gives water and no other residue contains two molecules of hydrogen to one molecule of oxygen.

This being so, and since we know the ratio by weight given above, namely, 8 to 1, we must conclude that the weight of one oxygen molecule is eight times that of two hydrogen molecules. Thus an oxygen atom is sixteen times as heavy as a hydrogen atom, and we say that the " atomic weight " of oxygen is 16 and the " molecular weight " of the hydrogen molecule is 2 and of the oxygen molecule 32.

This example shows how the physical theory of gases has a very definite bearing on chemistry, in that it determines the existence and relative weights of the atoms. Without the idea that the pressure directly indicates the number of molecules, this would be impossible.

We must content ourselves with this single example. In reality, our conviction that molecules are built up from atoms in this way rests on the vast number of chemical facts which are thereby given a clear and satisfactory explanation. In the following section we shall give a summary of those results of chemical research which have an important bearing on our purposes.

12. The " Mole " and Molecular Weight.

If N pieces of stone, each weighing 2 lb., weigh 2 tons, then N pieces, each weighing 32 lb., weigh

32 tons, for the piece of stone in the second case is sixteen times as heavy as in the first.

The same thing is true, of course, with molecules; the weights of N molecules of different substances are proportional to the weights of the molecules, i.e. to the " molecular weights ". If, then, we take 2 gm. of a substance of molecular weight 2, and 32 gm. of another substance of molecular weight 32, then these quantities must contain *exactly the same number of molecules*. A quantity of any substance equal to its molecular weight, in grammes, is called a *mole* of that substance, and we see that one mole of any substance contains exactly the same number of molecules as one mole of any other substance. The mole is a convenient quantity of matter, for if we deal in moles instead of grammes, we know that we are handling the same number of molecules of each substance.

The actual counting of the molecules need therefore only be done once for *one* gas, provided the molecular weights are determined once and for all. This is done by the chemists, who establish the relative weights of various substances which will combine with 1 gm. of hydrogen atoms to form a saturated compound. (If such does not exist we take some other substance, whose behaviour is already known, for comparison instead of hydrogen.) For purely practical reasons it has, moreover, been found preferable to measure these molecular weights relative not to 1 gm. of hydrogen atoms, but to 32 gm. of oxygen atoms, the latter substance being more convenient from the chemical point of view. Now the

ratio of the atomic weights of oxygen and hydrogen is not exactly 16 : 1 (as we said) but slightly less (16 : 1·008). Hence we say: relative to oxygen (O = 16), hydrogen (H) has the *atomic weight* 1·008.

As we shall see later (p. 255), the choice of oxygen as standard substance has turned out a very happy one.

What is the number of molecules in a mole? This number is usually called *Avogadro's number*, after the scientist who first developed the concept of the mole, and its value according to the latest determinations is $6·06 \times 10^{23}$. In order to get an idea of this colossal number, we may think of a cube containing this number of particles. Then along each edge there are $0·85 \times 10^8$ or 85 million particles (for this number when multiplied by itself twice gives $6·06 \times 10^{23}$). A " mole of peas "—yes, we can even speak of a mole of peas—would then fill a cube of edge 425 kilometres, if the peas were, as before, taken as $\frac{1}{2}$ cm. across.

As $6·06 \times 10^{23}$ atoms of hydrogen weigh exactly 1 gm., each hydrogen atom weighs $\dfrac{1}{6·06 \times 10^{23}}$ gm. or $\dfrac{10}{6·06} \times \dfrac{1}{10^{24}} = 1·65 \times 10^{-24}$ gm. This is the number by which the atomic or molecular weight of a substance must be multiplied in order to obtain the actual weight of an atom or molecule of the substance.

These are the numbers which enable us to convert the measures of ordinary life into those of atomic physics.

13. *The Periodic Table of the Elements.*

According to Chemistry, the incomprehensibly great number of substances of which dead and living matter is made up are compounds formed from a comparatively small number of elements.

There are 92 different kinds of atoms—the elements—and all molecules are combinations of some of these elemental atoms.

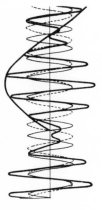

This analysis is one of the most wonderful achievements of the human mind. In essentials, it was complete at a time when the modern physical methods of atomic research did not exist or were still in their infancy. To separate and combine substances only the simplest methods were used, heating and cooling, dissolving, crystallizing, and filtering, a great part being played by personal judgment of such matters as look, smell, taste of the product aimed at. We might almost say that the utensils were those of a high-class cook. In addition there was a little physics; measurements of surface tension, electrical conductivity, and the like. Success, however, was mainly due to the age-long conviction that matter is at bottom simple and obeys simple laws—a purely theoretical idea! But there is no quest without faith that there is something to be found. There is no experimenting in the absence of a theory. The initial theory may be false, the desired goal unattainable. From the quest of the philosopher's stone and the alchemist's devices for making gold finally arose modern chemistry, which builds up the material universe out of 92 basic

TABLE I.—PERIODIC TABLE OF THE ELEMENTS

	I	II	III	IV	V	VI	VII	VIII
1	1 H 1·0078							2 He 4·002
2	3 Li 6·940	4 Be 9·02	5 B 10·82	6 C 12·00	7 N 14·008	8 O 16·0000	9 F 19·000	10 Ne 20·183
3	11 Na 22·997	12 Mg 24·32	13 Al 26·97	14 .i 28·06	15 P 31·02	16 S 32·06	17 Cl 35·457	18 A ← 39·944
4 ↑	19 K 39·096	20 Ca 40·08	21 Sc 45·10	22 Ti 47·90	23 V 50·95	24 Cr 52·01	25 Mn 54·93	26 Fe 55·84 27 Co ← → 28 Ni 58·94 58·69
	29 Cu 63·57	30 Zn 65·38	31 Ga 69·72	32 Ge 72·60	33 As 74·91	34 Se 78·96	35 Br 79·916	36 Kr 83·7
5	37 Rb 85·44	38 Sr 87·63	39 Y 88·92	40 Zr 91·22	41 Cb 93·3	42 Mo 96·0	43 Ma	44 Ru 101·7 45 Rh 102·91 46 Pd 106·7
	47 Ag 107·880	48 Cd 112·41	49 In 114·76	50 Sn 118·70	51 Sb 121·76	52 Te 127·61	← → 53 I 126·92	54 Xe 131·3
6	55 Cs 132·91	56 Ba 137·36	57 La 138·92	72 Hf 178·6	73 Ta 181·4	74 W 184·0	75 Re 186·31	76 Os 191·5 77 Ir 193·1 78 Pt 195·23
	79 Au 197·2	80 Hg 200·61	81 Tl 204·39	82 Pb 207·22	83 Bi 209·00	84 Po (210·0)	85 —	86 Rn 222
7	87 —	88 Ra 225·97	89 Ac (227)	90 Th ← → 232·12	91 Pa (231)	92 U 238·14		

85 As Astatin
94 Pu Plutonium

THE RARE EARTHS (to be inserted between 57 La and 72 Hf)

58 Ce 140·13	59 Pr 140·92	60 Nd 144·27	61 Il	62 Sm 150·43	63 Eu 152·0	64 Gd 157·3
65 Tb 159·2	66 Dy 162·46	67 Ho 163·5	68 Er 167·64	69 Tm 169·4	70 Yb 173·04	71 Lu 175·0

The numbers in front of the symbols of the elements denote the atomic numbers; the numbers underneath are the atomic weights. The latter are taken, with a few modifications, from the Report of the International Commission on Atomic Weights for 1932. The double arrow ← → indicates the places where the order of atomic weights and that of atomic numbers do not agree.

Since the original edition of this book new elements have been discovered; not only all gaps of the old table have been filled but also a number of "trans-uranium" elements have been found. These are:

43 Tc Technetium	61 Pm Promethium
87 Fr Francium	93 Np Neptunium
95 Am Americium	96 Cm Curium

" elements ". Indeed, the same spiritual urge has led research still farther—for who is going to believe that there really are as many as 92 ultimate units? But of this we shall speak later (Chapter V, p. 231).

In the first place, we can arrange these 92 elements for comparison, so as to get to know their properties.   Luckily we need not burden our memories too heavily. There are groups of elements of the same sort, which have similar chemical and physical properties. In the so-called *periodic table* (p. 51) these are arranged in vertical columns. Each element is denoted by its own symbol, its atomic weight (relative to $O = 16$) is given, and also a number increasing from left to right, the *atomic number*. In general the order is that of increasing atomic weight; there are, however, a few exceptions (denoted by ←→) which we shall discuss in Chapter IV (p. 203).

This table is itself the summary of a fundamental discovery (made by Lothar Meyer and Mendeléeff) which is as follows. We consider e.g. the group of elements lithium (Li), sodium (Na), potassium (K), rubidium (Rb), and cæsium (Cs), which stand in column I. They are all metals of similar physical and chemical properties; they are soft and readily combustible, they form salts of the same kind with chlorine or bromine, and so on. They are known as the alkali metals. If we pass from each of these elements to the next heaviest, i.e. from lithium to beryllium (Be), from sodium to magnesium (Mg), from potassium to calcium (Ca), from rubidium to strontium (Sr), from cæsium to barium (Ba), these latter elements also resemble one another in every respect, replace one another readily in chemical compounds, and so on. They are called the alkaline earth metals. If we pass from the alkali metals to the next lightest elements, we obtain a group of substances, all much alike, occupying the last vertical column. These agree in being gases and forming no compounds whatever—the so-called inert gases, helium (He), neon (Ne), argon (A), krypton (Kr), xenon (Xe), and radon (Rn).

This rule can be followed out, keeping strictly to the order given in the table. We see that in essentials there are eight vertical columns; that is, if we go up the atoms in order of increasing atomic weight we always come back to a related atom after eight steps.

This regularity, from which the periodic table gets its name, only holds exactly, however, at the beginning, for the two first horizontal rows. After that the true period is longer (18 steps), but the order of the table still lets the framework of the period 8 appear. Farther on, a number of elements appear to have been put in quite at random, the so-called rare earths, from cerium (Ce) to lutecium (Lu); and at one or two places, as we have already said, the order of atomic weight has been broken into in order to preserve the periodic grouping according to chemical similarity.

The series of elements should really be wound in a spiral on a cylinder (25). Then we should see more clearly that the inert gases come before the alkali metals, and that the former are preceded by a vertical

(25)

column of related substances, the halogens (salt-formers), fluorine (F), chlorine (Cl), bromine (Br), iodine (I); these are monovalent, that is, combine with just one hydrogen atom, forming the acids HF, HCl, &c.

Before the discovery of the periodic table the alchemists' idea of the relation and transmutability of the elements was a beautiful dream. Now it at once became a research programme. Proximity reveals genetic relationships, just as when the relations of form between organisms were interpreted by Darwin as meaning a succession in time, an evolution. The latest fashion among biologists is to throw doubt on the evolution theory, because we have " no proofs " of the transformation of species. O faithless generation! If physicists had done the same, they would have accepted the 92 elements and the periodic table as the gift of God—seeing that all attempts at transmutation had been in vain; modern atomic physics would never have arisen, and this book would have had to end here. But the genuine physicist believes obstinately in the simplicity and unity of Nature, despite any appearance to the contrary. Hence, for him, the periodic table was not a tombstone inscription, but a command to renew his inquiries.

CHAPTER II

Electrons and Ions

1. *The Question of the Ultimate Atom.*

IN the older atomic theory, which essentially depended on the facts of chemistry and the kinetic theory of gases, the word " atom " was used in its original sense of " the indivisible ". The atoms were regarded as little elastic balls, like billiard balls, each provided with a few little hooks (the " valencies " of the chemists), by means of which they could be linked to one another; and there were about 90 different kinds of them, out of which all substances were believed to be built up. But the material world is not so simple as this. Phenomena exist which do not fit into this scheme. For instance, it gives us no clue to the relation between light and matter. Light is emitted by glowing or burning bodies, absorbed by other bodies, transmitted by others more or less according to its wave-length, its velocity being altered in the process, and so on. Then there are the electrical and magnetic phenomena, also closely related to material bodies, and left quite out in the cold.

If all these phenomena are to be explained by so

simple an atomic theory, each little ball must be provided with many attributes. The balls can, however, differ only in their size, weight, and perhaps elastic properties. These differences do not give us anything like enough scope to explain the host of optical and electrical effects, to say nothing of the chemical properties that we must account for. We must therefore break the atom in pieces or dissect it. This is always the most effective way of finding out what a thing is made of. If we can succeed in doing so, we may hope that the connexion between the atoms, revealed in the periodic table of the elements, will also become intelligible.

When we dissect living tissue, we find an elemental living unit, the cell. Is there also an elemental unit, of which all atoms are built up, an ultimate atom?

More than a hundred years ago a physician named Prout asserted that the atom of hydrogen was this ultimate atom. If this were so, the weights of all atoms should be exact multiples of the weight of a hydrogen atom. In the first place, this is simply not true, as was proved by accurate experiments later, and as is shown by a glance at our table (p. 51). In the second place, the hydrogen atom is no more " simple " than any other atom, for it reacts to light, electricity, and magnetism in a complicated way, and can itself emit and absorb light of all colours. Thus there seems to be nothing in Prout's hypothesis; but we shall see that later it has been successfully re-stated in an altered form.

The ultimate atom, then, must be sought for

elsewhere. Some fifty years ago it was believed to have been found in the realm of electrical phenomena, in the atom of electricity or *electron*. Strictly speaking, however, this was no good either. The electron, it is true, is a kind of ultimate atom, but the 92 different atoms of matter cannot be explained merely by sticking electrons together. The problem is much more complicated: there are, as we shall see later (Chap. V), several other kinds of particle which contend for the honour of being the ultimate atom. In our dissection of the atom, therefore, we must go slowly, otherwise we shall do damage, like an anatomy student who makes too big a cut and destroys important organs. Just as he begins at the skin and removes layer after layer, we must proceed from the outside to the inside of the atom. We find that the outer shell of the atom *is* actually built up of atoms of electricity or electrons, and we shall describe how this has been established.

2. *Electrolytic Conduction.*

Before going on with our story it will be necessary to recall a few facts about electricity and magnetism. Everyone nowadays knows quite a lot about electricity, for he is using it in his house; he probably has a wireless set, possibly he owns a dynamo—he certainly does if he owns a car—and he can probably talk intelligently about volts, amperes, resistances, inductances, valves, triodes, and so on. We will begin, then, with something that practically everyone has seen and a great many people have actually handled and looked after, namely, an ordinary

wireless "low-voltage" accumulator, as it is called. This is a small celluloid cell containing an arrangement of lead plates in a dilute solution of sulphuric acid. When fully charged it will give us a supply of electricity at a voltage which the garage man describes as 2 volts.

The accumulator has two terminals, one marked red and the other marked black. The reason for this distinction in marking is to ensure that two or more accumulators are connected together in the right way. If we connect the two red terminals together with a copper wire and the two black terminals together with another copper wire, not much happens, but if we connect the red terminal of one to the black terminal of the other and vice versa, we shall certainly get a very bright spark when we complete the circuit, and, if we succeed in completing it without burning our fingers, we shall probably either melt the wire or, if we are not fortunate enough to do this, we shall ruin the two cells. What happens is very simple to understand. The meaning of the red and black terminals is that when we connect the red terminal to the black terminal by a wire, positive electricity will flow in the wire from the red terminal to the black terminal or (what comes to the same thing) negative electricity will flow from the black terminal to the red terminal, that is, in the reverse direction. If, on the other hand, we connect red to red, there is no tendency for electricity to flow in the wire.

Here we have supposed that there are two sorts of electricity; one sort we have called "positive"

and the other " negative ".   These two sorts of
electricity are known to exist, as can be proved
by the experiments we have all seen at school—
experiments that are made with glass rods, silk
rubbers, catskins, pieces of resin, and all the rest
of it, which we shall not recall here.   They are given
the names " positive " and " negative " as a matter
of convenience, to indicate that their properties are
rather like those of positive and negative numbers.
For instance, positive electricity repels positive electricity, negative electricity repels negative electricity,
but positive electricity attracts negative electricity.
This reminds us of our old rule in algebra: " Like
signs plus, unlike signs minus ", and this is, in fact,
the very reason why the signs " plus " and " minus "
are given to the two sorts of electricity, and why one
is called positive and the other negative.   Plus 5
and minus 5 added together come to nothing.   In
much the same way, a piece of matter to which we
give 5 units of positive electricity and then 5 units
of negative electricity shows no electrification at all.
The one charge has wiped out the other, so far as
outside effects are concerned.   We say that the piece
of matter is electrically " neutral ".

If now we connect a piece of electrical apparatus
across the red and black terminals of our accumulator, a current of electricity will flow, and this
current produces several very well-known effects.
For example, we all know that it will make a wire
red-hot and so give us light; it will make powerful
magnets out of pieces of soft iron, and so we have
electric motors.   But the current will do another
(E 969)

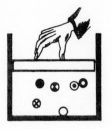

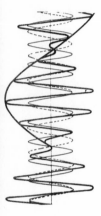

thing that is possibly not so well known, although very widely used in industry; it will decompose solutions of chemicals. The reader can very easily prove this for himself with his 2-volt accumulator. If he gets a jam-jar and half fills it with water into which he has put a pinch of salt, and then puts two metal rods into the solution, but not, of course, touching one another, he can connect this little apparatus (26) by copper wires to the red and black terminals of the accumulator. If he does this he will see that gases immediately begin to bubble off the two rods. The rods are called *electrodes*, from the Greek words *electron*, amber, and *hodos*, a path. The rod through which the positive current enters is called the *anode* (Greek *ana*, up), the other the *cathode* (*kata*, down). Gases, then, are seen to bubble off at the electrodes, and if the gases that come off each electrode are collected separately one is found to be hydrogen and the other oxygen, which we know, from chemistry, to be the constituents of water. Consequently our electric current has decomposed water for us.

The current will also decompose other substances. For instance, if we put into a jam-jar a weak solution of a blue substance called copper sulphate, and use two copper electrodes, we shall find that within a few minutes one of the electrodes has got a very definite coating of bright new copper on it, and the other electrode shows signs of having been nibbled at. If the two electrodes were carefully weighed it would be found that one had got heavier by the same amount as the other had got lighter, so that the electric current has simply conveyed copper from one electrode, where it has nibbled it off, to the other electrode, where it has deposited it. This copper has been handed from one electrode to the other through the solution of copper sulphate. The process is called " electrolysis ", and is the method used extensively in industry to prepare pure metals. It is also used in electroplating.

Now the point that is of special interest to us is the actual mechanism by which the current is passed through this solution of copper sulphate. In the first place, it may be said that if absolutely pure water is used the current will not pass at all; we say that the " resistance " of the solution is infinite. In order that the solution shall conduct the current at all, it must have some chemical substance dissolved in it. In one of the cases we mentioned above it was a pinch of salt, and in the other case it was copper sulphate. In neither case did the material collecting at the electrodes come from the added chemical substance. In the case of the copper sulphate, the copper that was added to one electrode had been nibbled off the other. It did not come out of the copper sulphate solution, for if it had the copper sulphate solution would be very much weaker at the end of the experiment than it was at the beginning, whereas its strength is the same, as chemists can prove.

By careful experiment with copper sulphate solution it has been shown that the longer we keep the current on, the greater the amount of copper that is deposited. If a current of water is running along

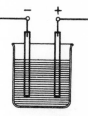

(26)

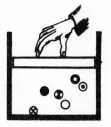

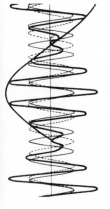

a pipe at a fixed speed, twice as much water runs along in two hours as in one hour. In just the same way, if the current in an electric circuit does not alter we send twice as much electricity round the circuit in two hours as in one hour. *But we also deposit twice as much copper in two hours as in one hour.* Consequently, it is clear that *the amount of copper* we deposit is simply proportional to *the amount of electricity* that goes through the cell.

This electricity has somehow to be handed through the cell from one electrode to the other. If one pound of copper takes part in handing one unit of electricity through the cell, we know, by the experiment just referred to, that two pounds of copper take part in handing two units of electricity through the cell. Now two pounds of copper contain just twice as many atoms of copper as one pound does. Is it not obvious, then, that the electricity must be handed through the cell in some way by the atoms of copper, so that one atom of copper carries along with it a perfectly *definite quantity of electricity*, neither more nor less? Or, in other words, does it not mean that there must be an atom of electricity? Otherwise each atom of copper might carry one sackful of electricity, as it were, but the sacks might not contain the same amount of electricity. Then there would be no necessary connexion between the number of atoms of copper deposited on one of the electrodes and the amount of electricity carried through the cell. The conclusion seems to be plain: each atom of copper, in passing through the solution and being deposited on one of the electrodes, carries with it 1, 2, 3, or whatever the number may be, of definite and indivisible atoms of electricity. It does not matter in the least if *each* copper atom carries two atoms of electricity—provided they all do the same thing—for so long as they all carry the same quantity of electricity, the number of carriers (atoms of copper) must be proportional to the amount of electricity carried.

This view has been worked out in detail and is found to account for all known facts of electrolysis and electrolytic conduction. We find that each atom of copper, silver, aluminium, or whatever the metal may be, carries a definite number of atoms of electricity. For example, a copper atom carries twice as many atoms of electricity as a hydrogen atom. So twice as many hydrogen atoms as copper atoms would be needed to carry the same amount of electricity. No material is known that needs more atoms to carry a given quantity of electricity than hydrogen does; we therefore conclude that hydrogen carries one atom of electricity. Here, then, we have a clear demonstration of the existence of atoms of electricity.

The charged atoms (or molecules) are called *ions* (Greek, wanderers). There are monovalent, divalent, &c., ions, corresponding to the number of atoms of electricity they carry.

The amount of electricity carried by 1 gm. of hydrogen is called *one faraday*, after the name of the great physicist who discovered the laws of electrolysis. One faraday is the *charge of the hydrogen ion per unit mass*, or the *specific charge* of the hydrogen

ion. As the number of atoms in 1 gm. of hydrogen is just 1 mole, i.e. 6×10^{23}, the same amount, that is, 1 faraday, of electricity is carried by 1 mole of any monovalent ion.

Of course scientists were not satisfied with this one set of conclusions, but tested the ionic hypothesis by a vast number of results. The theory of electrolytes is a whole science in itself; it deals with questions of how the resistance of the solution varies with the strength of the solution, the nature of the solution, and so on, and how all this is related to the other physical and chemical properties of the substances, and how the process is affected by temperature. But this study would lead us too far afield. It does not tell us anything new about what we are chiefly interested in, the atom of electricity. For we do not yet even know whether there are really two kinds of electrical atoms, positive and negative, or only one kind; in the latter case an ion of the other kind would be formed from a neutral atom, not by adding an atom of electricity but by taking one away.

How can we find out more about this?

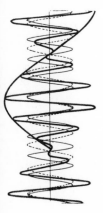

3. *Cathode Rays.*

Attempts had to be made to set the atom of electricity free and investigate it when detached from ordinary atoms. This was done in the researches of several physicists, e.g. Plücker, Hittorf, and Thomson, from the middle of last century onwards, on the conduction of electricity through rarefied gases. Long before this the Geissler tubes were

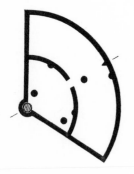

known. These tubes are filled with rarefied gas and light up with beautifully varied colours when an electric current is passed through them. A very complicated process goes on in the tube, a chaotic dance of charged atoms, which we learnt to control only by degrees. That we can now do this can be seen any day from the many-coloured electric signs used for advertising.

Though this process is of great importance, it is of more interest to the physicist to consider simpler phenomena. One way of advance consists in rarefying the gas in the tube more and more by pumping it out. Then the light becomes feebler and finally vanishes altogether. Nevertheless, a feeble current is still passing through the tube. And if we watch carefully we see something new: the part of the glass wall opposite one of the two electrodes (27) begins to shine with a green light! We said "one of the two electrodes". Looking more closely, we find that it is the cathode, the place where the positive current leaves the tube, i.e. which is negatively charged relative to the anode.

What is going on in the tube? Previously, when the gas was denser and the electricity "flowed" through the tube, we could see a luminous thread passing from anode to cathode. Now, at the lower pressures, something different happens: a ray-like effect comes from the cathode only. If small obstacles such as plates or wires are put into the tube, we clearly see their shadows in the green light on the glass wall. The radiation accordingly travels in straight lines; it is known as *cathode rays*. What

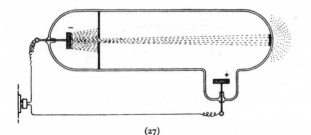

(27)

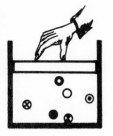

does it consist of? Is it a kind of light, that is, according to the usual idea, a wave motion? Or is it a rain of particles?

Gradually physicists became convinced that it was the latter. Above all, cathode rays carry electricity, negative electricity. This is indeed obvious from the fact that they fly out from the negative cathode (negative electricity of course repels other negative electricity); but the (negative) charge of the rays was also proved directly by catching them.

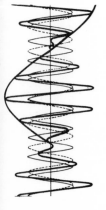

The physicists then tried to accelerate the rays, to retard them, or to deviate them from their paths. With these charged rays, this can be done much more easily than with the neutral molecular rays of which we spoke in Chapter I (p. 28). With molecules we have, in a sense, no point of attack, for, owing to the great velocity of the particles, gravity is not effective in deflecting them. Hence we had to produce an apparent deviation by rotating the beam apparatus. Electrically charged particles, however, are readily affected by electric and magnetic forces. There are numerous experimental methods of doing this. The details do not interest us. We only need to be clear about the principle.

4. *Electric and Magnetic Deflection.*

If we connect two parallel metal plates to the poles of a battery or of the lighting circuit (provided that this is direct current), one of the plates becomes positively charged, the other negatively (28). A small charged sphere brought between them will then be

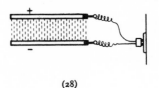

(28)

attracted by the plate which has the charge of the sign opposite to its own, and with the same force, no matter where it is between the plates. We say that between the plates there is a homogeneous *electric field*.

In this field the charged sphere falls just as a stone falls in the " gravitational field " of the earth, and if the little sphere is shot in from one side it will describe a parabola (29). There is, however, an essential difference between the two cases. We know that all bodies fall at the same rate—at least in a vacuum, where there is no air resistance. This is due to the fact that although a heavier body, owing to its inertia, resists deviation from its straight path more than a lighter body does, it is attracted more strongly by gravity, in the same ratio.

If, however, the force acting on the sphere is not gravity but the electric field, the state of affairs is different. The inertia or tendency to retain a straight path is determined by the weight (or better, the mass); but the deflecting force depends on the charge and is quite independent of the mass. If the electric force is increased, the body (i.e. the mass) remaining the same, the deflection is increased. Hence the curvature of the parabola, i.e. the amount of the final deflection, depends on the ratio of the charge and the mass, which is the quantity which on p. 63 we called the " specific charge ", in connexion with electrolytic ions.

Again, similar considerations apply if we shoot a charged body through a constant magnetic field, such as that which exists between the two poles (30)

(30)

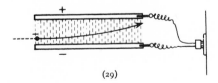

(29)

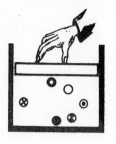

of an electromagnet. Of course the magnetic field acts on the path of the particle in quite a different way. An electrically charged particle at rest is subject to no force at all in a magnetic field. If the particle is moving, the magnetic field does not exert a force along its " lines of force " from pole to pole, but in a direction which is at right angles to them and also at right angles to the direction of motion of the particle (31). If the particle's flight is at right angles to the lines of force, it describes an arc of a circle (32). The curvature of this circle is again proportional to the specific charge.

The electric deflection and the magnetic deflection depend on the velocity in quite different ways. In the electrical case the deflection is inversely proportional to the square of the velocity, in the magnetic case to the velocity itself. Hence by combining the two types of deflection it is possible to separate the effect of the velocity from that of the specific charge and to determine each separately.

5. *The Specific Charge of the Electron.*

To state the actual numerical results of these experiments is of no great use except to those who are well acquainted with electrical units. We must already know some other quantity of the same kind if we are to judge whether a newly obtained number is large, small, normal, or abnormal.

Now in our case we have a comparison unit available, namely, the specific charge of electrolytic ions, in particular that of the (monovalent) hydrogen ion (one faraday; cf. p. 63).

When this number was compared with the specific charge of cathode rays as measured by deflection experiments, it was found that the latter is 1840 times as great—at least for sufficiently slow rays. (We shall discuss later what happens with faster rays.) Considering this from the point of view of the atom of electricity, we assume that the charge of the cathode-ray particle is the same as that of the monovalent electrolytic ion, but that its mass is 1840 times as small; for this makes the ratio $\dfrac{\text{charge}}{\text{mass}}$ just 1840 times as great.

This is so far satisfactory; for an ultimate atom must be lighter than all the atoms of matter which are to be constructed from it. Still, it is surprisingly light. Do 1840 of these " electrons " (as they were called) make up a hydrogen atom? No, the matter cannot be so simple as this, for the electrons are all negative, but the atom is neutral. Where does the positive electricity come from?

Before we relate how this was traced to the deepest recesses of the atom, we have still a good deal to say about electrons.

By applying a high electrical tension, the cathode rays can be accelerated. Later it was found that rays of the same sort occur among the radiations emitted by radium and other radioactive substances (p. 235). They are called β-rays and behave in every respect like fast cathode rays. Now if we measure not only the specific charge of the particles but also their velocity, we find that the specific charge decreases as the velocity increases. It is extremely improbable that

MAGNETIC FIELD

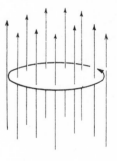

VELOCITY

FORCE

(31)

(32)

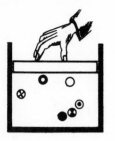

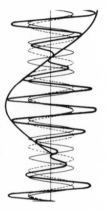

the charge itself should alter with the velocity. Hence there seems nothing for it but to assume that the mass—i.e. the resistance to acceleration—increases as the velocity increases: a very interesting result indeed, if true!

How can the mass of a body increase as we give it a greater and greater velocity? The reader will at once think, " Most likely the body is not alone, but has, attached to it, other unsuspected structures, so that if we drive it on faster, it must carry more and more of these structures with it."

Suppose you are trying to pull your motor-car out of the garage by hand. You put the engine out of gear, of course, otherwise you must pull much harder. The engine begins turning and you have to overcome the inertia and the friction of its parts. Of course you know that, but a negro from Central Africa unacquainted with the hidden interior of the car would be astonished—just as physicists were at the increase of resistance when electrons were accelerated to high velocities.

The physicist, however, looks deeply into Nature with the eye of his imagination; so he soon found out the hidden machinery, namely, the magnetic field produced when the motion begins.

Everybody knows the experiment with iron filings. A piece of paper is laid on top of a bar magnet and iron filings are scattered on it. Then these arrange themselves in lines running from pole to pole of the magnet, the so-called lines of force (33). These show how the magnetic force is distributed in direction and magnitude; where they run close

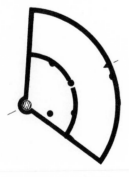

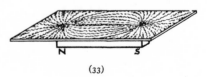

(33)

together, the force is strong; where they are far apart, the force is weak.

Now we imagine that the magnet is taken away, a hole is bored in the paper and a wire, through which we send an electric current, is stuck through.

(34)

If we again scatter iron filings on the paper, we obtain circular lines of magnetic force encircling the wire (34). This is the great discovery made by Oersted: every electric current is accompanied by a magnetic field (35).

This magnetic field represents a storing up of energy. If the current is increased, more magnetic energy is stored up, which must be supplied by the source of the current. True, this energy is not turned into heat, as by the resistance of the wire; still, it has to come from somewhere. Hence it looks as if there were a resistance to *change* of current. With a straight wire this effect is trifling; but if the wire is wound into a coil, so that every turn passes through the magnetic lines of force of the neighbouring turns, a powerful resistance to any change of current is produced, a so-called inductance. Anyone who has a wireless set is familiar with these inductance coils, which in a certain sense increase the inertia of the current, that is, its tendency to retain its strength unaltered.

(35)

Exactly the same thing happens when an electrically charged sphere moves. This moving charge, of course, is itself a sort of electric current, and is surrounded by magnetic lines of force (36). If the motion is uniform, the lines of force simply go with

(36)

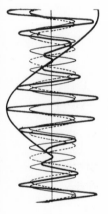

it. If, however, the velocity is increased or diminished, the magnetic field is increased or diminished, and this gives rise to an additional expenditure of force—just like putting in the clutch in the motor-car. It seems as if the sphere simply becomes heavier. In the case of a sphere which is not electrically charged this does not occur. Thus we have an apparent additional mass, which we shall call the *electromagnetic mass*. The main point is that this additional mass must actually increase as the velocity increases—as is observed in the case of fast cathode rays and β-rays.

The minuteness of the mass of the electron now led to the bold idea that *all* the mass of the electron just consists of this electromagnetic resistance to acceleration, and physicists spoke of *purely* electromagnetic mass! If, then, all atoms are constructed from electrons, we should have to assume that mass is always and everywhere an electromagnetic phenomenon, a sort of inductance. This, of course, would be an unheard-of simplification of the laws of Nature—but just the reverse of what the older physicists aimed at. They regarded mechanics as the fundamental science and sought to explain electrical and magnetic phenomena in a mechanical way, by means of invisible mechanism. Now we want to reduce the fundamental phenomenon of mechanics, inertial resistance, to electromagnetic laws.

Unfortunately things did not turn out so simply. The law of the change of mass with velocity deduced from this theory did not agree with observation;

besides, the theory was not complete—how, for example, do electron-spheres have their charges distributed? Are they on the surface, distributed uniformly through the interior, or how? The smaller the radius of the sphere, the greater does the electromagnetic mass become; thus the electron cannot be assumed to be a point, otherwise its mass would become infinitely great.

These difficulties led to the whole idea being given up when Einstein's theory of relativity showed a way of explaining the variability of mass without any special hypothesis. We must discuss this in greater detail.

6. *Some Remarks about Relativity.*

A friend of mine was once at a dinner-party and the lady next to him said: " Professor, do tell me in a few words what this theory of relativity really is." He replied: " Of course I will—provided you will let me tell you this little story first. I was going for a walk with a French friend and we got thirsty. By and by we came to a farm and I said: ' Let's buy a glass of milk here.' ' What's milk?' ' Oh, you don't know what milk is? It's the white liquid that——' ' What's white?' ' White? you don't know what that is either? Well, the swan——' ' What's swan?' ' Swan, the big bird with the bent neck.' ' What's bent?' ' Bent? Good heavens, don't you know that? Here, look at my arm: when I put it so, it's bent!' ' Oh, that's bent, is it? Now I know what milk is!' " Perhaps, like the lady, you do not want to hear any more about relativity. I

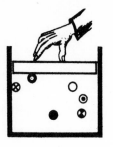

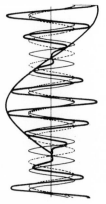

have, indeed, no intention of explaining that great theory to you here. I have already done so in a book you will find easy to understand. Here we are merely concerned with some of its results, which are of fundamental importance in connexion with atomic theory.

As those who go on cruises know, we can play ball games on board a liner just as on terra firma. This is what is meant by the mechanical principle of relativity, which goes back to Galileo and Newton. In abstract language it is as follows:

Mechanical processes, i.e. motions of all kinds, which take place in a room, say the so-called " frame of reference ", moving with uniform velocity in a straight line, go on exactly as they would in a room at rest. Really we are not entitled to speak of the latter as " at rest ", for as seen by people in other rooms, i.e. relative to other " bodies ", it is moving in the reverse direction, and it is impossible to tell (at least by mechanical means) which room is at rest. Relative to different frames of reference the laws of motion are the same, provided these frames of reference are moving uniformly in straight lines relative to one another. (The case is altered whenever accelerations, deviations from a straight course, or rotations occur: but we are not concerned with these at the moment.)

So far everything is in order and easy to understand. But of course there are other physical phenomena besides mechanical motions; for example, light, heat, electricity, and magnetism. We shall chiefly discuss light. Light does not take much

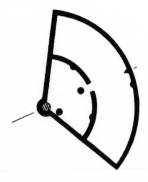

notice of material bodies. Once produced, it travels on through empty space from star to star.

Now light behaves like a vibration—we shall have much more to say about this later (p. 106). Where something is vibrating, there must, it is generally believed, be something to vibrate. Thus we are led to assume that even in so-called empty space there is something, the *ether*, whose vibrations are what we call " light ", and which, moreover, has the further duties of being responsible for the orderly behaviour of electric and magnetic phenomena. (We may remark in passing that light is an electromagnetic process.)

So it looks as if our principle that relative motions do not affect mechanical processes is incorrect. For the ether is obviously a body anchored in space, which can be said to be absolutely at rest.

The earth, however, is not at rest; it is moving at a rate of over 18 miles per second round the sun. It travels through the sea of ether in the same way as an aeroplane does through the air. Just as the aeroplane passengers observe a strong head-wind, so there must be an ether wind blowing past the earth. This is an important conclusion, for it can be tested experimentally!

The light-waves on the earth must be blown about by the ether wind. According as they travel with or against the direction of the ether wind, or at right angles to it, their velocity will be different. The effect is only a very minute one, for light travels about 10,000 times as fast as the earth travels round

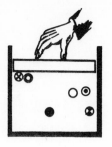

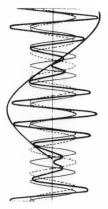

the sun. But first Michelson and then several others succeeded in raising the accuracy of measurement so high that less than a hundredth part of the effect should have been capable of detection.

Yet not a trace of it was found! Almost all physicists now agree on this point.

This, however, means that the principle of relativity applies to light also. The same could also be proved in the case of other electromagnetic phenomena.

This is really very astonishing. For the ether, which of course we could not do without (as the subject of the verb " to vibrate "), cannot be any substance of the kind with which we are familiar. Why should it be, anyway? " There are more things in heaven and earth . . . ," as Shakespeare said long ago.

The simplest explanation of the matter is obtained if (following Lorentz and Fitzgerald) we assume that the ether wind does not merely blow the light-waves away, but acts on all bodies. We have only to assume that all lengths are slightly contracted in the direction of motion; this just compensates the effect on the velocity of light.

At first glance, however, this " Lorentz-Fitzgerald contraction " is rather arbitrary. The state of affairs demanded a detailed examination, and this was carried out by Einstein. He considered two systems moving relatively to one another, say two planets, like the earth and Mars. On each of them physicists are supposed to be sitting in laboratories making similar experiments on light and other

phenomena. If there is no communication between them, they will have different units of length and time. But if receiving apparatus is so far developed on Mars that the Mars people can listen to the wireless stations on the earth, then they will hear the time signals. They will compare them with their own clocks, but, being well-trained astronomers, they will know that, in order to get simultaneity with the clocks on the earth they must take account of the time which the Hertzian waves take to travel from the earth to Mars. These waves travel with the speed of light, and as the distance between the two planets is known, this time difference can be calculated.

The Mars physicists will nevertheless be firmly convinced that their own time system is correct and that they know what they mean when they say that " two clocks at different positions on Mars indicate the *same* moment "; just as physicists on the earth believed they knew what they meant when they said " two events in London and New York are simultaneous "—until Einstein told them that they did not.

How can we know whether two events at different places on a planet are simultaneous? The best method, of course, is to use wireless time signals again. But then the velocity of the wave-signal comes in, and this depends on the accuracy with which times and lengths can be measured. Einstein showed how simultaneity can be defined *without* any knowledge about the wave velocity, except that it is the same in all directions. He simply assumed *three* stations instead of two, namely, A in New York,

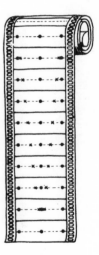

(37)

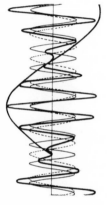

B in London, and C on a boat exactly midway between A and B. Then from C we send out a light signal in both directions at once; when it reaches A and B, we set the clocks there to the same time. In (37), p. 77, this time comparison is illustrated in a film. In the lowest picture we see the stations A, C, B; in the next, the two light-signals have been sent off and can be seen on either side of C. In the next picture they have travelled farther, and so on. The uppermost picture shows the arrival of the signal when the clocks in A and B are both set at, say, one o'clock.

Now the Mars astronomers have excellent telescopes and they can see the clocks in New York and in London! Will they agree that the two clocks are showing the same time? No, they will not. Figure (38) opposite shows the reason for this. It is also a film, only simplified. It again shows the stations A, B, C, but as they look from Mars. They are *moving* relatively to Mars, i.e. they are displaced in passing from one film picture to the next. Now the Mars physicists can construct the signal sent out in both directions in *their* film. Of course they have had *their* Michelson, though his name was different, but—this is the crux of the matter—he had made experiments which show that light *on Mars* travels in all directions with equal speed. That means that on the Mars film of the comparison of clocks on the earth, the two signals are travelling with equal speed to the left and to the right. But the left-hand station A is approaching the middle of the film, the right-hand station B is flying from

it. Therefore the left signal reaches A earlier than the other reaches B.

Now the Mars astronomers will declare that they cannot understand why the earth people have set the clocks at A and B in such a peculiar way; they show equal times when " really " the clock at A should be in advance of the clock at B. Suppose next that the earth people make the same observations on the clocks on Mars. They find them showing the same time when, from the standpoint of the observer on the earth, they should not show the same time at all.

Who is right? Who is wrong?

Einstein's answer is: neither is. They are both right and both wrong. Each planet, more generally each moving body, has its *own* system of time, and also of space; it can be shown that the Lorentz-Fitzgerald contraction is closely associated with the " relativity " of time. No planet can claim to have the *absolute* system of time and space. But if we know the velocity of the other planet *relative to our own* we can mathematically find out the readings of clocks and measuring rods as observed by people on the other planet. The formulæ used for this purpose are called the " Lorentz transformation ".

We need not say more about the numerous consequences of this theory of relativity. We must, however, consider the point we are interested in, namely, its effect on the laws of mechanics, more closely.

(38)

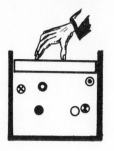

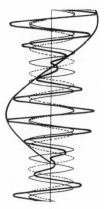

7. *Mass and Energy.*

The fundamental assumption in all the foregoing is that no signal faster than light exists. For if such did exist, it could be used for comparing clocks—and the whole theory of relativity would break down.

If this theory has any meaning, therefore, the universe must be so constructed that the velocity of light is the highest velocity which we can ever *observe.*

The laws of ordinary mechanics, however, are not compatible with this. Why should it not be possible to give a body an arbitrarily great acceleration by using powerful forces? Shells from large guns can already be made to travel faster than sound, so that they arrive before their own sound-waves— the victim is hit before he has time to hear the shot. Why should it not be possible to make electrons travel faster than light, by means of powerful electric fields?

If we stick to the relativity principle, we must alter ordinary mechanics. But nothing of the many established results of mechanics must be lost; remember the predictions of the planetary orbits by astronomers!

Now mechanics is not the last word of wisdom. We saw this already when we were discussing electro-magnetic mass. Electrical and magnetic phenomena may be even more fundamental than mechanics, so that mechanics may really be a " derived science ".

The beautiful thing is that the electromagnetic laws which we call Maxwell's equations carry the

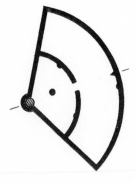

theory of relativity, as it were, on their very face. There is no difficulty about this; no alterations of any kind are needed, everything fits in excellently. For these equations have just the same form, no matter for what moving frame of reference they are stated, provided the system of space and time units which applies to that frame of reference is used. Thus we can immediately understand why " absolute " motion can never be determined by any electrical or magnetic observations (including optical observations); as Clerk-Maxwell predicted long ago.

We must not hesitate to regard ordinary mechanics—that wonderful structure of Galileo and Newton—as only approximately correct and to find out how it must be altered. This is not very difficult. Deviations are only to be anticipated in the case of very great velocities, near that of light. Newton's fundamental law is:

Time-rate of Change of Momentum = Force,

where by momentum we mean the product of mass and velocity (p. 10).

Now a body can have an acceleration, even if it is at rest; for it can begin to move, and so must be able to get up speed. We assume that Newton's law always holds if the body under consideration is *at rest*, so that it holds good in the frame of reference which is moving *with the same speed as the body*. In general the body will be at rest in this framework only momentarily; for the frame of reference is moving uniformly, whereas the body may be accelerated or retarded. Then at the next instant there is *another*

frame of reference moving with the same speed as the body; Newton's law must again apply in *this* frame of reference. We have now only to notice that in the two systems times and lengths are measured differently. The relations between the times and lengths then give the new law of motion. Formally the new law is just the same as Newton's, except that now the mass is no longer a fixed number peculiar to the body, but actually changes as the velocity changes.

And now a wonderful result comes out: if the velocity increases, the mass also increases, faster and faster, beyond all bounds, as the velocity approaches that of light.

In the first place, this is in exact agreement with the observations on cathode rays and β-rays. The relativistic mass-formula is actually confirmed excellently by quite accurate experiments.

In the second place, just try to make a body travel faster than light! It can't be done, for the faster the body goes, the heavier does it become. More and more force is needed to make it go faster, and in fact no finite force will raise its velocity to that of light.

Thus we have reached the result asserted: the velocity of light is the unattainable upper bound of all *observable* velocities. It is a natural unit of speed, which is actually used in physics; the high velocities of electrons and other particles are conveniently expressed as fractions of the velocity of light.

For low velocities (low compared with that of light), ordinary mechanics again applies. The mass becomes constant; we speak of the *rest-mass*.

This rest-mass, therefore, is the distinguishing mark of the body as regards mechanical properties. The theory of relativity has thrown fresh light on this idea also. For, according to this theory, *mass and energy are essentially the same.* If we multiply the mass of a body by the square of the velocity of light, we get its energy content, its capacity for doing work (a capacity which may, however, be fully utilizable in rare cases only).

There are many ways of explaining this theorem of Einstein's. The simplest is to consider what happens if a body is radiating light in one direction, e.g. a searchlight.

Now light carries energy in the form of " radiant heat ". Indeed, all life on the earth depends on the heat energy radiated to us by the sun. But light also carries *momentum*. The implications of this may be illustrated as follows.

When a man shoots with a gun he feels a recoil. The bullet flies forwards, so the gun must move backwards—otherwise the common centre of gravity could not remain at rest. For there is (p. 11) a fundamental law of mechanics, called the principle of the conservation of momentum, to the effect that the centre of gravity of a system of bodies which act only on one another does not change its state of rest or motion.

Quantitatively, it is the algebraic sum of all the momenta that does not alter. Before the shot, when the bullet is still in the gun, the momenta of the gun and of the bullet are both zero, and so also is their

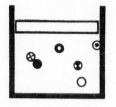

sum. After the shot, the bullet has a considerable momentum, and the gun must have acquired the same momentum with the opposite sign, that is, in the opposite direction, if the sum of the momenta is still to be zero.

The assertion that light possesses momentum accordingly means just this: emission of light in a certain direction leads to a recoil of the body emitting it.

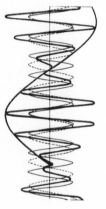

For example, we may take the searchlight and provide it with a shutter. If we open this for an instant, a beam of light rays shoots out. Then the searchlight is subject to a recoil like that of the gun.

That this is really so is not just a theory, but an experimental fact. True, the recoil is extremely feeble, and to detect it very light and fine suspended bodies must be used, very different from a big searchlight. It is easier to carry out the experiment the reverse way, by illuminating a light suspended disc with an intense source of light, and observing the deflection caused by the light (39) (p. 85). As the light is not like a single shot, but forms a continuous stream, we do not get a sudden impulse, but a uniform *light-pressure* as long as the light is shining. This pressure has been found to exist and to be in excellent agreement with the predictions of optical theory.

Light-pressure plays a considerable part in astronomy, owing to the very great intensity of the light emitted by the sun and stars. It is well known that the tails of comets always point away from the sun. They consist of very minute particles which are

blown back by the pressure of the light from the sun.

The principle of the conservation of momentum, then, applies to light also: a beam of light rays carries momentum with it, just as a bullet does. Hence a mass can also be ascribed to it; this is obtained by dividing the momentum by the velocity, that is, by the velocity of light.

Now the study of light-pressure has further shown, again in agreement with theory, that the energy (heat) transported by light must be divided by the velocity of light to give the momentum.

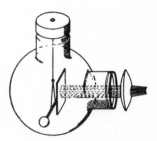

(39)

Accordingly, the mass of the beam of light rays is equal to its energy divided by the square of the velocity of light, i.e. $m = E/c^2$. Owing to the gigantic value of the velocity of light, the mass which "rides on the light" is very minute, for the factor $1/c^2$ is $1/(9 \times 10^{20})$, an extremely small number.

Yet there is no doubt that a body losing energy by radiating light does lose mass; but without necessarily losing particles. In the fixed stars, the amounts are quite large. Thus, in one year, the sun loses $1 \cdot 4 \times 10^{14}$ kg. ($1 \cdot 38 \times 10^{11}$ tons) by its radiation, though of course this is an inconsiderable amount compared with its mass of 2×10^{30} kg. Æons of time, then, will elapse before the sun radiates itself away.

If light is absorbed by a body, the body becomes warmer and simultaneously heavier by the mass corresponding to the light absorbed. Heat energy, therefore, has mass, just as light energy has.

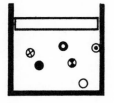

Every form of energy storage implies a storage of mass, no matter whether the energy be magnetic, chemical, or in any other form. Energy and mass are just different names for the same thing.

Each bit of matter is therefore a potential source of energy; if its mass could only be released, we should have immense supplies of energy at our disposal. For 1 gm. of mass is equivalent to $c^2(= 9 \times 10^{20})$ mechanical units of energy (so-called ergs); to obtain this energy from coal, we should have to burn nearly 3000 tons. Unfortunately, matter is not sufficiently obliging to set its mass and, so, its energy free.

Recently, however, cases have been found to occur where this transformation of matter into free energy actually takes place. We shall discuss this exciting subject in Chapter V. Here the restlessness of the universe reaches its highest pitch; the solid matter itself explodes, setting up a wild confusion of motion in its neighbourhood.

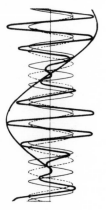

Meanwhile, however, we return to the electron. There is no doubt that, according to Einstein's theorem, its rest-mass also is nothing but energy— what sort of energy? Doubtless the electrical energy of the charge which it carries. Thus we are brought back to the idea of electromagnetic mass. The objections which have been raised against this idea are quashed—in my opinion—by Einstein's theorem that mass is energy. The electron, as a charged body, carries electrical energy of the right amount. Who can doubt that the mass is identical with this electrical energy?

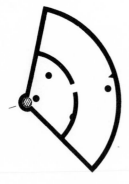

To charge up a small metal sphere, we have to do a certain amount of work, equal to the square of the charge (e) divided by the radius (r). This is the energy of the electric charge (e) on the sphere. If we assume that the same is true for the electron, by Einstein's theory we have

$$\frac{e^2}{r} = mc^2.$$

From this we obtain an estimate for the magnitude of the *radius of the electron*; we have

$$r = \frac{e^2}{mc^2} = 10^{-13} \text{ cm., approximately.}$$

But our objections? They arose in problems in which the laws of the electromagnetic field, investigated on the large scale, are applied to very minute particles. If they lead to contradictions and arbitrariness, so much the worse for the laws of the electromagnetic field! We have no ground for assuming that, in the inaccessibly small regions of the electron, the same Maxwell's laws hold as in large-scale laboratory experiments.

Starting from this conviction, I have worked out modifications of the laws, which avoid these difficulties. With the theory of relativity as basis, the new formulæ arise in a fairly natural way. They are identical with Maxwell's equations for all large-scale processes, and deviate from them only for very small dimensions. There, however, the difference is important; whereas, in the older form of the electromagnetic theory, the charged electron and the

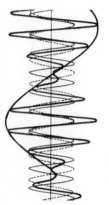

electromagnetic field surrounding it were two essentially different things, in the new theory they are united. There is only one field, but its properties are such that, at certain places, it may reach immense values without the store of energy blowing up. These places are the electrons. According to this theory, the electrons have a definite spatial structure, whose radius is determined by the formula given above and is about 10^{-13} cm. Although these considerations are satisfactory from the standpoint of classical field theory they cannot be easily adapted to quantum theory where the problem appears in a different light.

8. *The Determination of the Electronic Charge.*

We know the specific charge of the electron, i.e. the ratio $\frac{charge}{mass}$. What is the magnitude of the charge itself?

It is obtained very simply by using our knowledge of Avogadro's number (the number of particles in a mole). For 1 mole of electrons is the quantity of electricity which is called 1 faraday. If we divide this quantity of electricity by the number of particles in a mole, we have the charge on a single particle.

This method, however, is not very satisfactory. In the first place, it is inaccurate, as Avogadro's number is known only roughly from experiments with gases. In the second place, it would be nice to determine the charge on a single electron directly, and thus verify the atomic nature of electricity.

We should therefore like to construct a sort of electric balance, so sensitive that the addition of a

single electron would give a deflection. A bold
venture!

But it works, thanks to the comparatively large
forces to which a single electronic charge is subject
when placed in a strong electric field. The force is
equal to the product of the charge and the field-
strength, and though the charge may be small the
field-strength can be made large.

Of course the balance itself must be sufficiently
sensitive. It is usual to choose droplets floating
freely, generally of oil. The fine oil droplets are
brought between two metal plates, which are electrically charged, and the droplets are observed
through a telescope or microscope (40). If there is
no electric field the droplets fall by their own weight,
not with an acceleration like a large body, but at a
slow uniform rate. This is because the resistance of
the air to the fall of small spheres is relatively much
greater than for large. From this rate of fall the
radius of the droplets (assumed spherical) can be
calculated. Hence their weight is known.

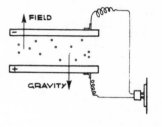

(40)

We now charge the particles. To do this we have
merely to illuminate the space between the plates for
a moment with light of short wave-length (ultra-
violet rays or X-rays). We shall discuss the electric
effect of this light, the *photo-electric effect*, in greater
detail later (p. 118). Here it is sufficient to know that
the light knocks electrons out of the molecules of
the air and thus makes the molecules into positive
ions. The electrons attach themselves to other
molecules of air, which thereby become negative
ions. And occasionally these ions attach themselves

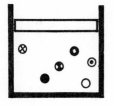

to a droplet of oil and give it an electric charge.

When the field is switched on we find that some drops are not affected by it at all, others fall more rapidly or more slowly than before, and some even rise in defiance of gravity. Now it is easy to measure the charge on a droplet: we may, for example, adjust the field-strength so that the droplet just remains floating in the field of the telescope. Then

Weight of Droplet = Charge × (Field-strength),

and if we know the weight from the rate of fall when there is no field, and measure the field-strength, we can find the charge, which is equal to

$$\frac{\text{Weight}}{\text{Field-strength}}$$

This is what Millikan did with a great many particles. He found charges of the most varying sizes, but among these there was a definite minimum charge, the smallest which was ever obtained. All other charges were exact multiples of this minimum charge.

The atomic nature of electricity is thus established, and the actual magnitude of the elementary charge (in the usual technical units) found. There is no point in stating the actual number. We must, however, attempt to give an idea of its magnitude by quoting effects which can readily be grasped.

If Millikan's plate apparatus is connected to the house lighting circuit àt 200 volts, how great is the force on one electron? That depends on the distance between the plates. If we imagine that the

plates are 1 mm. apart, the force is about as great as the weight of one three hundred millionth part of a milligram (3×10^{-9} mgm.). This seems very little—but if we choose an appropriate object for comparison, it is really enormously large. We may take the weight of the electron itself. As we know the charge and the ratio $\dfrac{\text{charge}}{\text{mass}}$, we can calculate the mass. We find that the electron weighs about 10^{-24} milligrams. The electric force is thus immensely greater, by the factor 3×10^{15} or three thousand billions.

Many readers will perhaps have wondered why we have said so little about gravitational forces. For these govern our whole destiny; they enable us to remain sticking to the earth, and they determine the path of the earth in its motion among the heavenly bodies. Must not gravity play an important part in the world of the atom?

We see that the answer is—no! True, it is very fashionable among present-day physicists to want to construct a unified theory which will combine the forces of gravitation and electromagnetism into one great whole. In my opinion, their attempts are along the wrong track, or at least premature. These effects are of quite different orders of magnitude, and occur under quite different conditions. In the realm of the atom gravity is completely masked by the electrical forces. It is only where these almost balance one another, as in bodies of large dimensions, which are a neutral collection of elementary electrical particles, that gravity becomes noticeable. Perhaps

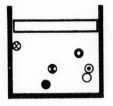

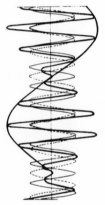

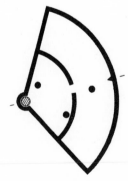

it is a residue due to incomplete compensation. But at present speculations about it are premature, for the true laws of electromagnetism are probably not yet known with sufficient accuracy.

Now we go back to the charge on the electron. Another way of getting an idea of its magnitude is this: how close together must two equal electronic charges come, if they are to attract or repel one another (according to their signs) with a force equal to the mechanical unit 1 dyne, which, as we know, is nearly equal to the weight of one milligram? This distance is found to be about 5×10^{-10} cm., that is, less than the radius of the atom. But 1 mgm. is of course a very arbitrary mass. To get a better idea of the forces, we think of one electron as fixed and the other shot straight at it. As they repel one another, they cannot approach indefinitely, but there must be a turning-point. How quickly must the electron be shot in order to come within atomic distance (41) (1 Å. $= 10^{-8}$ cm.)?

The answer is, with a velocity of 10^8 cm./sec. or 1000 Km./sec., i.e. 1/300 of the velocity of light. If we used heavier particles instead of electrons, the projectile would approach the fixed charge more closely still, the velocity being the same. We shall make use of this result later (p. 171) in estimating magnitudes in the interior of the atom.

The electronic charge has been measured very accurately; its true value certainly does not differ from the experimental value by more than 1/1000 of that value. Hence it may be used to obtain a more accurate value for Avogadro's number;

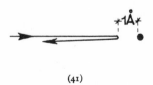

(41)

we have of course merely to divide the faraday by the charge of the electron.   There are yet other methods for determining the electronic charge.   For example, as we shall soon see (p. 97), we can count the particles in the rays emitted by radioactive substances.   If at the same time we measure the total charge carried by the rays, we obtain the charge on a single particle by division.

These and other quite independent methods agree in giving the same value, and thus confirm our conviction that the electron—the atom of electricity —actually exists.

9. *Gaseous Ions.*

We now know quite a lot about the electron. The next question is, what part does it play in the building-up of matter?

The electrons of the cathode rays come out of the metal of the cathode. Hence presumably the metal is full of electrons, and it is these that are responsible for the high electrical conductivity of the metal.

But we said previously that electrons can be torn out of air molecules, and indeed out of atoms and molecules of every kind, by irradiating them with light of short wave-length.   This phenomenon is known as the *photo-electric effect*. It occurs with the so-called ultra-violet light, whose wave-length is shorter than that of the visible violet light, but which cannot be detected by the eye.   It is shown still better by X-rays, which are really just light of very much shorter wave-length.   In many gases the torn-out electrons move about freely for some

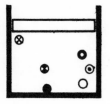

time. If an electric current is made to pass through tubes containing traces of these gases, it is carried chiefly by the free electrons, which move much faster than the positive ionic residues. They rush with increasing velocity to the anode, reaching very high velocities, even when the applied electrical forces are not high. The speed is so great that the atoms of the gas become luminous. Of this kind of light we shall speak later (p. 130). The process is put to practical use in electrical advertising signs, which emit pure neon- or argon-light and require but little current.

In most gases, however, the electrons knocked out of the atoms are immediately caught by other atoms, giving rise to negative ions. An irradiated gas then conducts electricity, just like an electrolyte. The current is carried by the ions, the positive ions going to the cathode and the negative to the anode. The velocities and charges of the ions can be measured, but we need not consider the details here.

Besides light, there are other means of splitting off electrons or "ionizing" molecules. Any kind of fast-moving particle with an electric charge will do. Hitherto we have spoken only of the cathode rays, which are fast-moving electrons. These can be set free from their prison in the evacuated tube by Lenard's method of inserting a window of very thin metal foil in the tube. The rays pass right through the metal foil and can then be observed in air or other gases, and their penetration, ionizing effect, and other properties can be investigated.

We also mentioned the naturally-occurring

electron rays, the β-rays. There are also rays of positive particles, naturally-occurring ones from radioactive substances and artificial ones produced in evacuated tubes. The ionization produced by these various types of rays may be stronger or weaker, but its nature is always the same. It is in every case, for every atom, a splitting-off of electrons.

This is extremely important. If you are eating an apple and biting off bit after bit, you have the same stuff in your mouth all the time, apple. Just so with atoms; you can remove bit after bit, electrons all the time.

These, as we know, are negative electric charges. The atom as a whole is neutral. Where, then, is the corresponding positive electricity? An apple has a core, which you finally reach. Just in the same way, if you keep peeling off layers of the atom you finally come to a (positive) core or nucleus—in the case of a molecule, several nuclei.

Before discussing these we have still much to say about electrons, and shall first describe the practical applications of what we have learned. For every new result leads to the construction of new apparatus, and this again to further new advances.

10. *Measuring and Counting Particles: Amplification: Making Particles Visible.*

The pieces of apparatus which we shall now describe are the most important of all the tools of modern physics. It is by means of them that the results which we have discussed above have been made physical certainties.

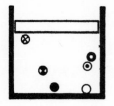

In our pictures we have shown atoms as dots or little balls and have represented their motions. But the actual atoms we have never seen yet, nor perceived them in any other way as individual entities.

It is a triumph of modern experimental technique to have made this possible in some degree. Well, then, you may say, why have we had to read all the pages that have gone before? Would it not have been simpler to produce the direct modern methods at once? Surely it would have been a saving of words, time and trouble?

No, it would not have done. You would not have understood the pictures and descriptions which now you will understand without further explanation (at least, I hope you will). There is nothing which is really so immediately illuminating as raw sense-impressions. Anyone can tell instantly whether a drink tastes sweet or bitter. The question where the sweet taste comes from, however—perhaps from sugar mixed with the drink—requires observation and investigation. Every experiment, even the simplest and most direct experiment, requires for its interpretation a knowledge of certain concepts and a capacity to arrange sense-impressions in order. These concepts had first to be formed, and their use learned, by the analysis of natural phenomena. I think, too, that this troublesome working-out of the truth actually has a great charm of its own, like that felt by a mountain climber who has scrambled up the steep slopes to the summit and scorns the people coming up by the mountain railway.

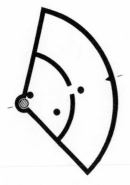

Now to business. We want to use the electrical

properties of gases to investigate rays of all kinds. Our simplest instrument is the *ionization chamber*, a metal vessel with an insulated electrode. Between this electrode and the wall of the vessel an electrical tension (so many volts) is applied. The air (or other gas filling the vessel) is a non-conductor in its natural state, provided the electrical tension is not so high as to make a spark pass. If ionizing rays are allowed to enter the vessel, pairs of ions are formed, the gas becomes a conductor, and if a small electrical tension is applied, a current begins to flow, which can be read off on an instrument (an ammeter).

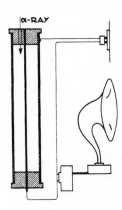

(42)

This apparatus enables us to compare and to measure the total effects of rays. We can, for example, find out by how much a radiation is weakened on passing through a layer of a substance, say a thin sheet of metal, and we can thus draw conclusions about the nature of the radiation. These conclusions, however, are still merely indirect.

Much more can be done with an instrument known as the *Geiger-Müller counting apparatus* (42). For this actually enables us to count the individual particles in material radiation—electrons or ions. It again consists of a metal tube and an electrode inside it, which is in the form of a wire covered with a thin layer of badly-conducting substance (oxide layer). A considerable electrical tension is applied between the wire and the wall of the metal tube, until the insulating layer of air is on the point of breaking down. The gas is then in an unstable condition; the electrical forces are very great, particularly in

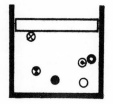

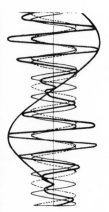

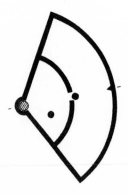

the immediate neighbourhood of the wire. These forces give a considerable velocity to any electron which happens to be present, until it hits the next air molecule, almost succeeding in ionizing it. If this does happen, another electron will be split off from the air molecule and will begin to move about. The two electrons will fly on, their number doubling at each (successful) collision. This avalanche-like increase (43) (p. 99) leads to the catastrophe; a spontaneous electric charge takes place and—if it is powerful enough—a spark is seen.

We accordingly adjust the electrical tension so that this catastrophe just fails to occur (here the oxide layer is useful).

Then if an electron or charged ion rushes through the space, the few hundred or thousand electrons to which it gives rise will suffice to bring about the discharge, and a current-impulse flows through the apparatus. By suitably adjusting the external circuit (inserting high resistances) we can arrange that the current is immediately broken off again when the particle has passed through the chamber. Then the apparatus is ready to record the passage of another particle; actually to *record* it, for the current-impulses can not only be made visible as deflections of a pointer-instrument, but can be led into an automatic counting mechanism. This can be done very conveniently by means of an amplifying valve of the very same kind as is used in wireless sets.

So now we have come round again to the starting-point of this chapter. I said that to-day everybody knows something about amplifying valves, triodes, and the like. Now, however, we are in a somewhat better position for understanding how these valves work.

The *amplifying valve* is a cathode-ray tube with a few modifications. In the first place, a relatively large current must pass through it. The number of electrons produced by a moderate electrical tension is too trifling to carry a powerful current. Hence we assist the process by means of heat. A glowing metal emits electrons spontaneously and copiously. This is not difficult to understand. As we said before, the high electrical conductivity of metals indicates that the electrons in metal are free—or nearly free— to move about, like the molecules of a gas. Raising the temperature means raising their velocity. The incandescence means that the electrons have high velocities in the metal, with which they run up against the surface of the metal from the inside and over- come the forces which in the cold state suffice to imprison them.

Thus we have a glowing cathode, that is, a thin wire heated electrically. Then we have a cold anode, and between them a third electrode in the form of a network, termed the " grid ". Hence the name *triode* for the valve as a whole (44)·(p. 101).

A 50–120 volt battery is connected between the cathode and the anode, but a current cannot flow in this circuit (indicated in the figure by ——·——) unless the electrons coming from the hot cathode do actually reach the anode.

This, in the first instance, is prevented by applying

(43)

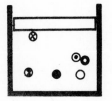

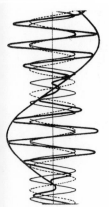

an opposing electrical tension between the cathode and the grid, so that the majority of the electrons just fail to reach the grid. Further, the alternating current which is to be amplified is transferred to this auxiliary circuit by a transforming coil. If this current (shown by a heavy dotted line) is in the same direction as the auxiliary electrical tension (dots and dashes), no current flows in the main circuit, any more than it did before. But whenever it is in the opposite direction to the auxiliary grid field, all of a sudden a great number of electrons reach the grid, pass through the holes in it, and thus complete the main circuit between the cathode and anode.

Quite small fluctuations in the electrical tension of the grid (relative to the cathode) are therefore able to produce enormous fluctuations of the current in the main circuit (for the source of current in this circuit can be made as powerful as we please). These fluctuations are transferred by a transformer coil to the wire which is to carry away the amplified current.

As an illustration, we may cite the mechanical machines in which powerful effects are produced by a tiny lever. An example from commerce, however, seems more convincing to me, that of customs at the frontier of a country. Think of a country importing sugar, say. A duty is imposed on this to control the import of sugar, to protect home industry. The duty corresponds to the electrical tension between grid and cathode, the quantity of sugar imported to the main current. A slight increase in the duty means that the import of sugar ceases to be profitable and the supply stops entirely. A small decrease in

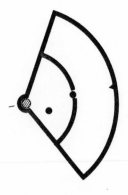

the duty again enables the importer to make a small profit on every pound of sugar he sells, and immediately the goods pour in, in unlimited quantity.

The same happens in the case of the valve; a slight decrease in the relative electrical tension of the grid lets every single electron leaving the cathode get through to the anode.

This is the principle of the amplifying valve. Its practical realization demanded an uncommonly high mastery of the art of the glass-blower, the production of extremely high vacua, and a knowledge of the properties of metals. These, however, we are not concerned with here.

By means of these amplifiers we can now magnify the feeble current-impulses of the Geiger-Muller counting apparatus as much as we like; we can make them audible as crackles in a loud-speaker or convey them to a counting mechanism from which the number of impulses, i.e. the number of electrons shot in, can be read off directly.

Anything we can count has an individual character. Anyone who has heard the crackles in the loud-speaker when a radium preparation is held near the counting-tube can no longer doubt that the radiation from radium is discontinuous.

It would, however, be very nice to *see* the atoms as well as *hear* them. Even this wish can be fulfilled. There is a method—not, it is true, for seeing the individual particles, but their collisions with a solid body—which dates back further than the counting method just described. A certain crystal,

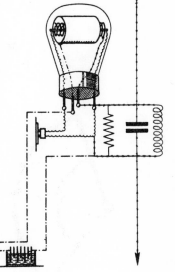

(44)

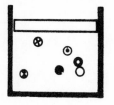

zinc sulphide, gives a flash of light when it is struck by a fast particle from a radium preparation (a so-called α-particle, in reality a helium ion (p. 235)). In a darkened room it is possible with a not very powerful magnifying-glass to see the individual hits as little points of light, so-called *scintillations*, which can easily be counted if the eye is suitably rested. Anyone who has a watch with a luminous dial can verify this; for the figures are painted over with zinc sulphide powder mixed with traces of a radio-active substance. To the naked eye the figures seem to be feebly luminous; but a magnifying-glass shows that the light is really intermittent.

The most elegant apparatus for making the whole tracks of particles visible, however, is the *Wilson chamber*. It is really no wonder that this was invented in a country like Great Britain, where thick fog is so frequent and troublesome a phenomenon. Why are London and Manchester foggier—and dirtier—cities than Berlin and Munich? Yes, dirtier too; for fog and dirt go together. The reason is the British love for the open coal fire with its incomplete combustion. Innumerable chimneys vomit forth soot day after day (45), and soot gives rise to fog.

The connexion between them is this: air can mix with water vapour, but not in any quantities. There is a maximum of humidity; we then say that the air is saturated with water vapour. This point depends on the height of the barometer, i.e. the pressure of the atmosphere. If the barometer suddenly falls, the quantity of water which the air

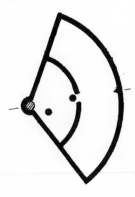

can retain, as vapour, falls, and the surplus condenses into droplets.

(45)

If, however, the pressure is lowered cautiously and the air is pure, condensation can be avoided. There is indeed more water vapour than the air can really hold, but the formation of droplets does not occur—in a sense because the water molecules do not know where to begin lumping themselves together. This task is made much easier for them, however, if there is any dust or soot floating in the air. At once the water molecules rush to the surface of these granules, cover them, and speedily form droplets with a nucleus of soot. There you have your fog! And down it sinks, covering everything with a greasy black layer of soot.

In Berlin and Munich, where the coal is burned thoroughly in central-heating plants, the air remains clearer and purer; on the other hand, the comfort of the crackling open coal fire is not known there (46).

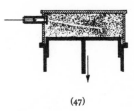

(46)

Now it has been found that not only dust and soot, but also charged molecules (ions) form excellent nuclei for drops to form on. The electric forces radiating from the ion obviously attract the water molecules.

This fact is utilized in the Wilson chamber to make the paths of particles visible. We have a chamber full of air saturated with water vapour, closed by a piston. If the piston is suddenly pulled out, lowering the pressure in the chamber, the vapour is super-saturated, but remains for a short time in that state, as there are no dust particles present to act as nuclei for droplet formation. If

(47)

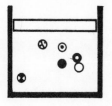

we now shoot particles into the chamber, these form pairs of ions along their path, which act as condensation nuclei and collect a layer of water droplets on their surfaces; the paths of the particles are rendered visible (47) (p. 103) as fine tracks of fog! These can be photographed; it is best to do this stereoscopically, that is, two cameras are used simultaneously from different directions, giving two pictures from which the positions of the tracks in space can be established.

Each kind of particle gives its own characteristic track; Plate I(*a*), facing p. 242, shows an electron and an α-particle, which can be distinguished at the very first glance: the light electron has a fine short zigzag track, the heavy α-particle a long, smooth, straight track.

The Wilson chamber method banishes the last doubt that matter consists of very fine particles, and we can now proceed with a clear conscience to study the part played by electrons in the make-up of atoms, to hunt for the positive constituents of atoms, and so on. This is what scientists actually did. Here, however, we shall not follow the historical path. For it led into a jungle of difficulties, from which they could extricate themselves only by slow degrees. The chief of these difficulties was the fundamental discovery that phenomena exist in which the very rays which we have just recognized with complete certainty as a rain of particles behave like *waves*!

Here we have a definite and horrible contradiction! Physicists, who were climbing down step by step so rapidly into the interior of the restless

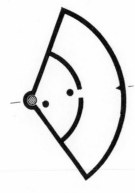

universe, were suddenly confronted by a blank wall which cut off any further advance.

Before we penetrate deeper, therefore, we must demolish this wall, get rid of this contradiction—so far as possible. First, however, we must consider waves and their properties a little more closely.

CHAPTER III

Waves and Particles

1. *Light-waves and Interference.*

LIGHT is the most important messenger bringing us news from the outside world. What does it really tell us? We think we actually see *things*, their outlines and colours. In reality the light merely reports this: " I come from such and such a direction, vibrate with such and such an intensity and such and such a frequency, and I have entirely forgotten what happened to me on the journey on which I set out just after my birth and which ends here on your retina with my death." Everything else, such as our perception of coloured objects, is not like a newspaper reporter's " copy ", but is an unconscious combination by the editorial department (the brain) of thousands and thousands of these reporter's messages, depending on impressions derived from all the senses taken together.

Most people find this journalistic combination so fascinating that they scarcely pay any attention to the reporter's skill. The physicist, however, finds these very reports particularly attractive. He does not combine them unconsciously, but, on the contrary, deliberately sets out to analyse them, using considerable ingenuity and cunningly devised apparatus. Then they tell him quite a different story, one of a restless universe of atoms, governed by strange laws.

Light itself forms part of the restlessness in the world of the very small. Even where there are no atoms, in empty interstellar space, for example, there are light-rays coming from the stars and moving in all directions. And near the stars, which of course are glowing suns, like our sun, light vies with the atoms, rushing on in a wild dance.

As we have already said, it is usual to regard light as a wave motion, each wave-length corresponding to a definite colour. At the time of the great Newton, this idea was not accepted without contradiction. Newton himself preferred the hypothesis that light is a rain of particles (" corpuscles ") given out by the body emitting the light. He could not understand how the wave theory, which had by then been devised by Huyghens, could explain the propagation of light-rays in straight lines (i.e. the fact that shadows have sharp boundaries), which is immediately obvious on the basis of the corpuscular theory. Newton made great discoveries in optics: in particular, he succeeded in splitting up white sunlight into the colours of the rainbow by means of a prism; as we should say to-day, he produced its *spectrum*. But in the question " wave or particle?" the scientific world came down on the side of Huyghens, for very forcible reasons, which we must now consider.

In the first place, light does not by any means

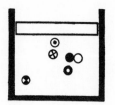

travel in a straight line in all circumstances. If we make a very small hole in a cardboard or metal screen and put a lamp behind it, the hole, when viewed from the back of the screen in any direction, looks like a small luminous point. This cannot be explained by Newton's theory, but can by Huyghens': the opening on which the light-waves are falling acts like a secondary centre from which the waves spread out in spheres. This experiment and all the following experiments can be imitated with water waves, although the latter are of course not waves in space but waves on the surface of a liquid. If a stone is thrown into water it produces a system of circular waves (Plate I(*b*), facing p. 242), whereas in space (in the case of light) the waves are spherical, starting from a luminous particle (i.e. a vibrating particle). At great distances the waves have a nearly straight front. A straight wave can also be produced by laying a stick in the water and moving it periodically. Plate I(*c*) shows a straight wave impinging on a board with a hole in it; behind the board we see a circular wave starting from the hole. This exactly corresponds to the optical experiment mentioned above.

Two trains of waves can pass through one another without affecting one another; that is, once they have again separated, each one goes on as if it had never met the other. This can be seen very well from a lake steamer; the waves made by the steamer pass on through the waves already present on the surface of the lake (48).

This rule is called the *principle of superposition*.

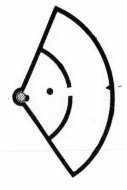

Exactly the same is true of light. If it were not so, how could we see at all? If I look in a certain direction, the waves which reach my eye are crossed by innumerable other waves on the way, but are not disturbed by them.

What happens, then, at the point where the two wave-trains are superposed?

Plate I(d) (facing p. 242) shows what happens with water waves. Two wooden balls floating some distance apart on water are set in periodic motion by a string. Then each of them sends out a system of circular waves, and these give rise to a curious wave-pattern; at certain points the waves reinforce one another, while at other points they destroy one another and the water remains still.

This is a fundamental phenomenon of wave motion, which is called *interference*. At points where a wave-crest of one system meets a wave-crest of the other, they reinforce each other; but where a wave-crest of one system meets a wave-trough of the other, they weaken each other; if the crest and trough are equally well marked, they cancel one another completely.

Applied to light, this means that light + light does not always give more light, but may in certain circumstances give darkness. Now is this really so?

About 120 years ago physics went through a period which was just as rich in important discoveries as the present. Among these the first observation of the interference of light by Young and the development of the wave theory by Fresnel are perhaps the greatest.

(48)

Young made two narrow slits in a screen and allowed light to pass through these and fall on another screen at some distance. Then he actually saw dark and bright bands (fringes) alternately, just as the wave theory predicts. Similar bands appear when fine obstacles such as hairs are observed (Plate II(d), facing p. 246). For our purposes, however, it is more convenient to think of two long narrow slits, on which monochromatic light, that is, light of one definite wave-length, is allowed to fall, the light passing through being caught on a screen. Then it is easy to calculate and mark points on the screen where one wave is exactly a whole wave-length behind the other; there a bright band will be found, and each pair of bright bands will be separated by a dark band (49). (The screen with the interference bands on it is really perpendicular to the plane of the paper.)

If the distance between successive bands is measured and the distances between the two slits and between the slits and the screen are known, the wave-length of the light can be calculated from a simple geometrical argument.

This construction also shows that the interference bands become narrower if the slits are moved farther apart, and conversely. The distance between the slits and the distance between the bands are inversely proportional to one another.

If the light falling on the slits is not monochromatic, but consists of a mixture of light of different wave-lengths, like white light, each wave-length gives its own system of bands, and the super-

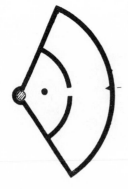

position of these gives the eye the impression of coloured and more or less blurred bands.

By using a great number of parallel slits it is possible to get each wave-length to produce quite narrow bands with large dark interspaces. Then the bands of different wave-lengths (colours) will not be superposed and mixed up, but will lie side by side. Only the centre band will contain all the wave-lengths at once, i.e. will appear white (in white .light). The very next band will have all the colours side by side, that is, will form a complete spectrum (the so-called spectrum of the first order). Each succeeding band is also drawn out into a spectrum, called the spectrum of the second order, third order, and so on.

Apparatus of this kind is· called an *interferometer*, and for many purposes is much better than the prisms which Newton used to split up light into a spectrum; for one thing, an interferometer can be used to determine wave-lengths. Instead of slits in a screen it is usual to have a metal mirror ruled with fine parallel lines; this is called a " grating ". Some gratings have as many as 2000 lines to the millimetre. A grating gives a whole series of spectra in addition to the undecomposed central image.

The complete agreement of the wave theory with the facts forms a convincing proof of the correctness of Huyghens' hypothesis.

How, then, are we to explain the rectilinear propagation of light which puzzled Newton so?

This is shown by Plate I(*e*) and (*f*) (facing p. 242), which are photographs from above of water waves

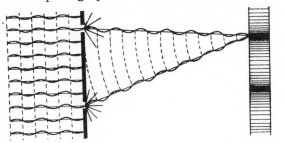

(49)

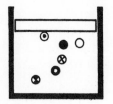

passing through a slit. If the slit is large compared with the wave-length, the edge of the shadow is comparatively sharp. If the width of the slit is reduced, the wave motion extends beyond the boundary of the shadow; we have what is called " diffraction ". If the slit is made quite small, a shadow-boundary is no longer to be seen, and the wave emerges from the slit (secondary source) as a circular wave-system (in space, a spherical wave-system). This can also be explained theoretically. A large opening may be imagined to be broken up into portions, each of them as long as a wave-length. Each such portion sends waves to a point beyond the shadow-boundary; half of them reach there as a wave-crest, half as a wave-trough, and they therefore cancel one another. At the edges, however, the compensation is incomplete, so that the shadow-boundary is not quite sharp and in certain circumstances may even exhibit narrow bands. If the slit is made narrower, the waves do not cancel one another so exactly and an extensive wave-train is produced. In the optical case we should see the image of the slit on the screen widening, its breadth being inversely proportional to the breadth of the slit. We shall come back to this point later (p. 160).

2. *Invisible Light.*

All kinds of light and everything of the nature of light have been examined by apparatus of the type just discussed.

Visible light covers only about one octave, speaking in musical terms; that is, the longest light-waves,

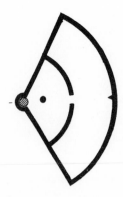

which appear red, are about twice as long as the shortest light-waves, which appear violet. The visible spectrum lies in the region of wave-lengths of 1/2000 mm.; more accurately, it consists of wave-lengths from $3 \cdot 6 \times 10^{-5}$ cm. to $7 \cdot 8 \times 10^{-5}$ cm.

It follows from this that the vibrations are exceedingly rapid. For these tiny waves (say of wave-length 5×10^{-5} cm.) advance 3×10^{10} cm. in a second (for that is the velocity of light). As one centimetre contains $\dfrac{1}{5 \times 10^{-5}} = 2 \times 10^4$ such waves, $2 \times 10^4 \times 3 \times 10^{10} = 6 \times 10^{14}$ waves pass a definite spot in a second. This is the " frequency " (ν) of light from the middle of the visible spectrum.

The light emitted by the sun and other sources of white light contains waves both shorter and longer than this.

If we hold a sensitive thermometer in the spectrum, it indicates heat; if we place it beyond the last visible colour on the red side, the thermometer rises still farther. Thus there is " infra-red " light, which can be detected by its heating effects and in other respects behaves just as visible light does. On the violet side of the spectrum the presence of " ultra-violet " light can be detected by means of the photographic plate. Thus the spectrum extends to an immense distance on either side, so that the region to which the eye is sensitive is quite a small portion of the spectrum as now known, so small that in making a diagram of the spectrum (50) we have to adopt a special device. We cannot plot the wavelengths themselves, for there would be no room on

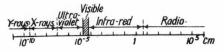

(50)

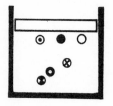

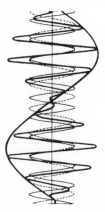

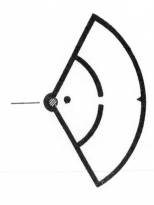

the paper; instead, we plot their orders of magnitude, going up in powers of 10.

Beyond the ultra-violet rays on the short-wave side are the X-rays. Their discovery by Röntgen in 1896 marks the beginning of the new radiation physics. The older among us still remember the almost alarming astonishment with which we were seized when we saw the photograph of the bones of a living man's hand. The young generation accepts this marvel and many others as obvious. The miracle depends on the extraordinary penetrating power of X-rays. It was not easy to prove that they are waves of the same kind as light; the wave-length is so small that sufficiently fine gratings could not be made.

Here, however, Nature came to the scientists' help, for crystals are wonderfully regular arrangements of atoms. Figure (51) opposite shows the structure of rock-salt, which consists of sodium and chlorine atoms arranged alternately in a very simple way. Von Laue had the bright idea of using crystals as gratings for X-rays. True, they are not gratings of lines, but gratings in space, so that the phenomenon becomes more complicated; but W. H. Bragg and his son W. L. Bragg made use of it to develop not only the spectroscopy of X-rays, but also the investigation of the grating-structure, or so-called *lattice-structure*, of crystals. As a matter of fact, it is now possible to produce X-ray spectra with artificial gratings also.

The radioactive substances emit not only the corpuscular rays which we have already mentioned (p. 102) but also a light-radiation, the so-called γ-rays, resembling X-rays. Finally, in the cosmic radiation which we shall discuss later (p. 261) we have light of extremely short wave-length.

Now for the long waves. Here the nature of light is revealed, so far as we can say so in physics at present. The question is, what is it that is vibrating?

This question led physicists to assume the existence of the *ether*. A hundred years ago the ether was regarded as an elastic body, something like a jelly, but much stiffer and lighter, so that it could vibrate extremely rapidly. But a great many phenomena, culminating in the Michelson experiment and the theory of relativity, showed that the ether must be something very different from ordinary terrestrial substances.

Now an ether is also required for electricity and magnetism; for these phenomena can also travel through a vacuum. Formerly physicists had no hesitation in filling space with all manner of different ethers. But the urge to unification of the concept of the universe, which is a powerful spur to research, forbade them to be content with this. Faraday's experiments led Maxwell to the idea that light is just a vibration of electromagnetic force. He predicted that circuits in which alternating electric currents are rapidly pulsating must emit waves of electric and magnetic force, and these were actually detected by Hertz. They are the waves which are used in wireless telegraphy to carry messages, and in radio for the entertainment of the public.

A wireless antenna, then, is just like an atom sending out light-waves, only greatly magnified.

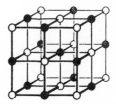

(51)

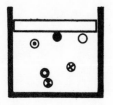

Film III shows how the radiation arises. Here the antenna is a short, straight wire. In this wire a current is oscillating to and fro; that is, at a certain instant one end of the wire is positively charged, the other negatively charged. At the latter, electrons are heaped up, at the former there is a shortage of electrons. Now the electrons flow away from the negative end, but overshoot the mark, so that at the next instant the charges are distributed in the reverse way. Thus the charge oscillates backwards and forwards, until its energy is used up by friction and radiation. How is the radiation produced? At the instant when the two charges are separate, there are two electric poles there (we call the whole thing a *dipole*), and outside the wire there is an electric field running from the negative pole to the positive pole; this field may be visualized by means of the lines of force. When the charges cancel one another, the dipole is destroyed and the lines of force detach themselves from the antenna and move outwards into space. During the existence of the reversed dipole a new system of lines of force is built up, but in the reverse direction, and so on, as the film shows.

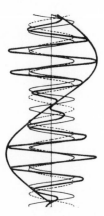

Actually there is not only the oscillating electric field, but also a magnetic field, whose lines of force surround the antenna. We cannot, however, go into details of this.

We can produce Hertzian waves as short as heat waves and just as long as we like. Their physical identity with light-waves as regards velocity, interference, reflection, and many other properties, has

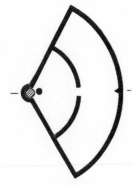

been proved by countless experiments. Hence we must assume that atoms emitting light are also little oscillating dipoles. As we already know that the atoms contain electrons in their outer layers, this idea raises no difficulties.

Up to the beginning of the present century this electromagnetic theory of light seemed firmly established, when suddenly a catastrophe shook the theory to its very foundations.

3. *Light Quanta.*

This catastrophe was not brought about all of a sudden by an unforeseen discovery, but arose in very much the same way as political changes do.

The great revolution in physics began with the work of a single man, Max Planck. By extremely careful experiments he showed that in certain phenomena of heat radiation the observed facts cannot be reconciled with the hitherto accepted laws of mechanics and optical theory. Once he was absolutely convinced of this, he sought to make a very trifling modification in these laws, which would bring them into agreement with the facts. In 1900 he asserted that it is necessary to assume that the emission and the absorption of light take place in *quanta*—" atoms ", we may say—not in arbitrarily small amounts (as was possible according to the wave theory). And further, for light of a definite colour the amount of energy (E) taken in or given out by an atom is proportional to the frequency (ν), so that

$$E = h \times \nu.$$

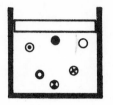

The number h is now known as Planck's Constant. This h is extremely minute; if the energy is measured in mechanical units (ergs) and the frequency is taken as the number of vibrations per second, $h = 6.5 \times 10^{-27}$ erg-seconds. As we have seen, visible light vibrates about 6×10^{14} times a second; Planck's quantum of energy for this kind of light, therefore, is only $6.5 \times 10^{-27} \times 6 \times 10^{14} = 4 \times 10^{-12}$ ergs, an extremely small amount. Yet this very minute discontinuity of the process, assumed by Planck, had dramatic consequences! Five years later Einstein came forward and declared that Planck had said far too little. According to him, discontinuity does not merely occur in the emission and absorption of light; no, light itself by no means consists of smooth waves, but is quite discontinuous or " quantized ": in short, it behaves like a rain of particles: *photons* or *light quanta*.

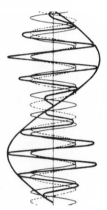

This is Newton's old hypothesis again, but now armed with quite new experimental facts, above all, the observations of the *photo-electric effect*.

This we have already mentioned (p. 93). If light of short wave-length falls on matter, it knocks out electrons. The process has been investigated by means of *photo-electric cells*; these are evacuated glass tubes with a coating of metal (e.g. sodium) on their inner surface, and provided with a quartz window which lets through ultra-violet light (52). These cells are used for a variety of practical purposes, e.g. in talking films and in television apparatus. There are also photo-electric instruments for measuring the intensity of light, which photographers find useful in estimating exposures.

Physicists have accurately investigated the connexion between the number and velocity of the electrons emitted and the properties of the light. If the intensity of the light is increased, the current of electrons emitted by the metal becomes stronger, but not, as one might expect, because the electrons are more accelerated by the stronger vibrations and fly out of the metal more quickly—no, the velocity remains the same so long as the colour or, more accurately, the wave-length of the light remains the same; it is the quantity of electrons emitted that increases. If, however, the wave-length of the light is altered, the velocity of the emitted electrons is altered in accordance with Planck's law.

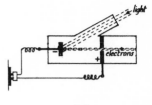

(52)

Here it must be noticed that the energy which an electron possesses in the interior of the metal is not completely available as kinetic energy when it leaves the metal; for the electron is bound in some way in the interior of the metal—otherwise, of course, the metal would spit out electrons spontaneously. We may say that the energy level of the electron lies deeper in the metal than it does outside, so that the electron must be raised (as regards energy) in order to get free (53). The electrons are like people in a tube station who can move about freely though in a restricted space; but if they want to get up to the ground level, they have to go up in the lift or stand on the escalator. The motive power of the lift or escalator thereby does a certain amount of work, which depends on the difference of level. In the

(53)

I. Gas Molecules

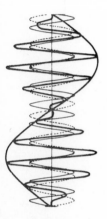

IV. Group Velocity

II. Molecular Velocities

very same way the electron has to be lifted out of the interior of the metal, and a certain amount of work (A) is required to do this. Then it is found by experiment that the ratio

$$\frac{E + A}{\nu}$$

always has the same value.

Accurate measurements have shown that, for all substances and in all circumstances, this value agrees with the value of h which Planck deduced from quite different experiments.

From the point of view of the wave theory all this is quite unintelligible. Why, on this theory, should there be so close a connexion between energy and frequency? The contradiction becomes still more glaring if we observe the photo-electric effect with very small metal particles. These are brought between charged metal plates as in Millikan's oil-droplet method (p. 90), and the instant of charging is revealed by a sudden change in the rate of fall of the particle. If the phenomenon took place in accordance with the wave theory, an electron of definite velocity could never be driven out until the metal particle had absorbed an amount of wave-energy corresponding to this velocity, plus the work required to separate the electron from the atom. But this is by no means the case: emission of electrons is occasionally, though rarely, observed to take place as soon as the light is switched on, long before the minute particle could possibly have accumulated enough energy.

Einstein asserted that this puzzling behaviour would at once become intelligible if light were regarded as a rain of particles (photons), whose energy, following Planck, is $h\nu$. But we see at once that such particles cannot be real in the sense of having mass; for if they had mass and moved with the speed of light, their energy would be infinite, as the theory of relativity tells us (p. 82). A photon falling on an atom can give up its energy to an electron instantly and knock it out of the atom. The quantity of electrons knocked out by light is accordingly proportional to the quantity of photons, and the energy of the electrons (less the work required to separate them from the atom) is proportional to the frequency.

At first physicists were extremely sceptical about this idea; for the wave theory seemed excellently confirmed by countless experiments and measurements. But gradually there accumulated a host of experimental facts which immediately became intelligible on Einstein's hypothesis, whereas the wave theory could not do anything with them; and these were mostly phenomena in which light was transformed into other forms of energy, or conversely.

We shall discuss two important groups of phenomena. The first are to some extent intelligible on Planck's original theory, in which the emission and the absorption of light take place in quantum jumps. The second group, however, raise the problem of the quantization of free radiation.

III. HERTZIAN OSCILLATOR

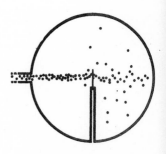

V. SCATTERING OF α-PARTICLES

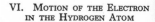

VI. MOTION OF THE ELECTRON IN THE HYDROGEN ATOM

VII. ROTATION OF THE ORBITAL PLANE

4. *The Spectral Lines of Gases.*

It was the radiation from hot solid bodies that led Planck to the idea of quanta of energy. But a solid body is in every way a complicated object, a conglomeration of countless atoms crowded together into a small space. Hence its radiation is a whirl of countless differing waves, and it is almost a miracle that from so complicated a phenomenon it was found possible to deduce a fundamental law. There are, however, simpler substances, in which the atoms or molecules are so loosely arranged that they scarcely affect one another, namely, the gases, to which we always come back. If we wish to find out the properties of atoms we shall accordingly be well advised to consider the radiation of gases.

Imagine a schoolboy who has never heard an opera before. He and his parents happen to arrive too late and must wait for a while outside the closed doors. He hears the music going on, the singing and theatrical effects, a wild hubbub of clamorous noises. If the father says, " Listen, there's the tenor singing ", or " Now that's a lovely violin solo ", the boy will gape at him in bewilderment. The physicists were in a similar position when they first investigated the radiation from solid bodies—a sort of optical music, played by countless unknown instruments behind closed doors.

A gas, however, roughly corresponds to the group of " first violins " by themselves; instruments all of one kind, though still of unknown structure and played behind closed doors.

At last the doors of the opera-house are opened, the novice sees the singers and orchestra and soon understands how they co-operate; some day he may have an opportunity of handling a violin or a flute and finding out how their sounds are produced.

The physicist can only see his optical orchestra with the eye of his imagination. He is very pleased if he can pick out a set of instruments of the same kind (the atoms of gases) and study their part by itself.

The spectrum of an incandescent solid, produced by a grating or prism, contains all the wave-lengths of the visible region; it is continuous. On the other hand, the spectrum of a gas consists of separate bright lines, often very numerous. This means that the atoms of gases give rise to vibrations of definite frequencies, which are propagated as waves of the corresponding wave-length.

This is nothing very wonderful; the same occurs in acoustics. Thus a piano string has a definite proper note (the so-called fundamental tone) associated with *one* frequency. By the exercise of a little ingenuity, however, it can be made to produce higher notes (overtones) in addition to the fundamental tone; now the string vibrates twice, three, four, . . . times as fast, giving the octave, octave and fifth, &c. Hold down middle C on the piano so that the note does not sound, the key merely being depressed. Then strike the C in the bass clef shortly and sharply. The C which is being held down can then be heard sounding on quite distinctly. For the lower C contains not only its own funda-

(E969)

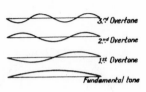

3.ʳᵈ Overtone

2.ⁿᵈ Overtone

1.ˢᵗ Overtone

Fundamental tone

(54)

mental tone, but its octave, middle C, as its first overtone; the air excited by the string of the lower C makes the string corresponding to middle C vibrate, and if the tuning is good, this string responds so strongly (by resonance) that it can still be heard when the lower C string is damped down by its key being allowed to rise. The same thing can be demonstrated with the higher overtones.

The overtones of the string correspond to vibrations in which the string divides itself up into vibrating portions separated by so-called *nodes*, points which are at rest: one node for the first overtone, two for the second, and so on (54).

Some other instruments, such as bells, have overtones whose frequencies are *not* exactly twice, thrice, . . . that of the fundamental, so that their vibrations are much more complicated.

How far does the analogy apply to optics? Perhaps the spectral lines are just the fundamental tone and overtones of the atoms, which are vibrating like elastic bodies?

This idea is such a natural one that it has been thoroughly investigated, but with entire lack of success. For the lines in the spectra of gases have an appearance which cannot arise from the vibrations of any elastic body.

At first sight the lines appear to be in such confusion that they seemingly follow no rule (see, for example, the illustration of part of the spectrum of iron * (Plate II(*a*), facing p. 246).

* This photograph is reproduced by permission of Adam Hilger, Ltd., from whom copies of the original can be purchased.

In the end, however, it was found possible to bring order into this confusion in the case of hydrogen, which, as the lightest atom, is presumably the simplest. If a Geissler tube is filled with hydrogen, it gives a line spectrum (Plate II(*b*)) whose lines follow a simple rule. Balmer discovered this rule, which we shall have to discuss later (p. 177). Then it was found possible to obtain similar laws for atoms which have one loose electron, that is, behave as monovalent positive ions in solution; these are the alkali metals already mentioned, lithium, sodium, potassium, &c. The lines of a gas of this kind form so-called *series*, whose regularity is at once obvious when one succeeds in observing them alone or picking them out of the confusion of other lines. Plate II(*c*) (potassium spectrum) shows a series of this kind. It immediately suggests overtones, and yet there is a fundamental difference: the lines move closer and closer together and are clearly piling up towards a limit beyond which no more lines of the series are to be found. Acoustic overtones never do this, for the frequencies of the elastic vibrations of any body or system of bodies form a series which has no end, just as in the very simple case of the piano string where the series is the succession of numbers. 1, 2, 3, . . . to infinity.

In the case of the spectral lines, however, another simple law has been discovered, called Ritz's *combination principle*, after its discoverer. This is best seen in groups of lines which obviously are closely related, so-called " doublets " or " triplets ". Thus

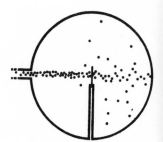

in incandescent sodium vapour there occur doublets
with the wave-lengths

$$\begin{cases} 8194\cdot94 \times 10^{-8} \text{ cm.} \\ 8183\cdot31 \times 10^{-8} \text{ cm.} \end{cases}$$

and

$$\begin{cases} 5888\cdot26 \times 10^{-8} \text{ cm.} \\ 5882\cdot90 \times 10^{-8} \text{ cm.} \end{cases}$$

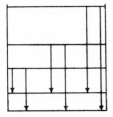

(55)

If we calculate the frequencies of these, by dividing
the velocity of light, 3×10^{10} cm./sec., by the wave-
lengths, we obtain exactly the same difference each
time, and this very difference also occurs with a
great number of other pairs of lines. This relation
can be illustrated by the accompanying diagram (55),
in which the frequencies are represented by vertical
lines, in the case of a doublet always starting from
the same height. The two end-points are placed
so that the equality of the differences is obvious:
the ends of the lines lie on two horizontal straight
lines. Analogous diagrams can be made for triplets,
quartets, &c.

From this the following general rule naturally
results. For a definite atom it is possible to draw a
number of horizontal straight lines, or *levels*, and
we can then represent all the frequencies of its
spectral lines as vertical distances between these
levels. The numbers determining the positions of
the levels are of course fewer than the frequencies.
In the above example there are five levels and six
frequencies, but often the number of lines is considerably higher than the number of levels. By no
means all the combinations of levels in pairs, how-

ever, appear as lines in the spectrum. We shall give diagrams of levels later; the reader may like to glance at these, say at that of hydrogen (p. 178), in order to get an idea of the arrangement of the levels in different series and of the spectral lines shown as the lines joining the levels.

The positions of the levels can be given on a frequency scale, in which the absolute value, that is, the number associated with the lowest level, is quite undetermined. These level-numbers are also called *terms*, and the combination principle is then stated as follows: each atom has a number of terms, a term-diagram, and every spectral line is the difference of two terms; the frequency of the line is equal to the upper term less the lower term. This principle of Ritz's has proved unfailingly correct in the analysis of innumerable spectra, in the visible region and also in the infra-red and ultra-violet, not to mention X-rays and γ-rays. But what· does it *mean?*

5. *Bohr's Theory of Spectral Lines.*

In 1913 Bohr recognized the connexion between Ritz's principle and Planck's quantum theory, first in the case of hydrogen and later quite generally. He did not, like most physicists, spend his time investigating and hesitating about the correctness of Planck's assertion that the emission and absorption of light take place in jumps of energy. Instead of the statement

$$\text{Frequency} = \frac{\text{Energy emitted}}{h},$$

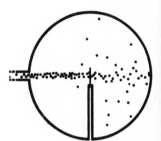

Bohr wrote the equivalent statement

$$\text{Frequency} = \frac{\text{Energy of Atom before Emission}}{h}$$

$$- \frac{\text{Energy of Atom after Emission}}{h}.$$

Now we have merely to combine this with Ritz's principle in order to reach the conclusion that

$$\text{Term} = \frac{\text{Energy of Atom}}{h},$$

or

$$\text{Energy of Atom} = \text{Term} \times h. \enspace \bullet$$

The meaning of this formula is that each atom exists in a number of states, which Bohr called *stationary states*, each of which has a definite energy. The energy of the atom can change, by absorption or emission of light; then the frequency of the light is equal to the energy-difference of two stationary states, divided by h.

Of course Bohr's assumption contradicts ordinary mechanics. If the atoms were subject to the ordinary laws of mechanics, energy could be pumped into or out of the atom in as small quantities as we pleased. The fact that Bohr's theory denies this is, however, a great advantage; for quite a primitive argument shows that the atoms cannot be mechanical systems, like the sun and its planets. At normal atmospheric pressure an atom collides with other atoms over 100 million times a second. Yet its spectral lines are sharp, and their frequency is unaltered. Imagine, on the other hand, what would happen if our planetary system should meet a fixed star, say Sirius, merely passing near it—not to mention actually colliding with it! All the planets would be hurled out of their orbits, all their periods of revolution would be altered—and that at a single encounter with another mechanical system.

The atoms, however, exhibit colossal stability and resistance to collisions, such as are never exhibited by the ordinary mechanical systems which we are familiar with in our immediate neighbourhood.

If Bohr's ideas are right, this stability at once becomes intelligible. For the atoms in their natural state will be in the lowest and most stable energy level (ground-state); the minimum energy which will throw them out of it is that corresponding to the distance between the lowest level and the second level. From the term-diagram, however, we see that this is very large compared with the thermal energy of the atoms flying around. Hence collisions between atoms are not sufficient to throw an atom out of the ground-state or, as we say, to excite it.

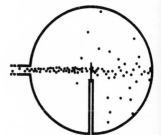

In addition, Bohr's theory immediately explains the fact that the emission spectrum of a gas contains many more lines than the absorption spectrum. We get an absorption spectrum if we look at an incandescent body, which is emitting light of all frequencies, through the cool gas. Then certain lines appear dark against the bright continuous spectrum. The Fraunhofer lines in the sun's spectrum (Plate III(*a*), facing p. 250) form an example: they are due to the absorption of the light from the central body

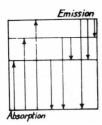

Emission

Absorption

(56)

(57)

of the sun in the cooler outer layers of the sun's atmosphere.

The cool, unexcited gas is in the ground-state; hence only those lines can be absorbed which arise from transitions from the ground-state to higher states. On the other hand, an atom which is excited in a higher state can emit many more lines, not only those which lead back directly to the ground-state, but also those which lead to any of the intervening states; and from these the atom may return by further steps to the ground-state (56).

This idea of *excitation* being necessary before an atom can radiate proved extremely fruitful in explaining the phenomena of light emission and in the development of electric lighting (neon lamps, &c.); for it gave information about the conditions under which certain lines were to be expected.

Thus, for example, atoms can be made to emit light by shooting electrons at them. By varying the velocity of the electrons it is possible to supply more or less energy to the atom. These experiments, first carried out by Franck and Hertz, yielded conclusive proof of the correctness of Bohr's theory. For they showed that as long as the energy of the bombarding electrons is smaller than that corresponding to the first stage of excitation, nothing happens at all; the atom does not emit light, and the electrons rebound from the atoms without loss of energy. As soon as the electronic energy rises a little above this level, however, the first line appears. We have the wonderful phenomenon, previously unknown, of the one-line spectrum. As the energy

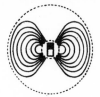

of the bombarding electrons is increased further, the lines appear in the order given by the term-diagram (Plate II(*e*), facing p. 246). We can even measure the energy lost by the electrons on collision and compare it with the term-value: complete agreement has always resulted—except that as a rule the optical measurements are much more accurate than the electrical ones.

Bohr's interpretation of the terms as energy levels gets rid of the arbitrariness of the zero point, which we mentioned above (p. 127). The levels become closer and closer as we go up; they approach a limit. This can clearly be seen from the diagrams of the lines of a series (see, for example, Plate II(*b*), (*c*)) and the term-diagram (57). Beyond the limit there is a region of continuous absorption. The meaning of this is clear. If more and more energy is supplied to the atom, an electron will finally fly out. The limit of the series of terms corresponds to the work required to separate the electron from the atom, that is, the energy required to ionize the atom. As a rule this is taken as the zero point, and the stages of excitation below this are counted negative; then the ground-state has the greatest negative energy value. The continuous absorption spectrum beyond the limit means that as the electron flies off it can carry any amount of kinetic energy away with it.

Even before Bohr put forward his theory an immense amount of spectroscopic data had been accumulated. These data had been collected because all physicists felt that here a great secret lay hid.

By one stroke of genius the secret was laid bare, and order was brought into the chaos of numbers and observations.

This, however, was only the first step; the next problem was, how do the energy levels of the atom arise? Ordinary "classical" mechanics tells us nothing. New quantum mechanics must be devised. Here again Bohr himself made the first advance. In Chapter IV we shall come back to this. It is very remarkable that the whole structure of a new mechanics, known as *quantum mechanics*, was gradually built up merely by better and better interpretation of facts about spectra. In 1925 Heisenberg put forward a decisive idea; this was seized on by Jordan and myself, who worked out the appropriate mathematics, the so-called *matrix mechanics*.

You may wonder how this came about. A student occasionally goes to lectures about abstruse subjects just for fun and speedily forgets all about them. This is what happened to me with a lecture on higher algebra, of which I recollected little more than the word "matrix" and a few simple theorems about these matrices. But that sufficed. A little playing about with Heisenberg's physical formula showed the connexion. Then it was an easy matter to refresh my memory and apply the results. This form of quantum mechanics, which was also brought to a high degree of perfection by Dirac quite independently, is not only the earliest form of quantum mechanics, but perhaps also the most fundamental; but it is so mathematical and abstract that it cannot

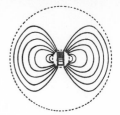

be made intelligible without the use of mathematics.

We shall accordingly drop this whole train of thought for the time being and return to it later. Meanwhile we shall turn to other phenomena which provide an easier approach to the new mechanics of the atom.

6. *A Photo-electric Game of Billiards.*

We discussed the scattering of light when we attempted to explain the nature of the blue of the sky and the red of the sunset (p. 38). There we used the illustration of a ship in rough water, which is made to pitch and roll by a train of waves and gives rise to secondary waves.

We can now understand the process better from the point of view of the electromagnetic theory of light. The atom or molecule is a structure containing electrons; somewhere in its interior it must contain a positive charge balancing them. If the atom comes into an electric field, the negative electrons are pulled to one side and the positive nucleus to the other; thus the two electric charges are separated, giving rise to a dipole, just as in the charged Hertzian antenna (p. 116).

Now a light-wave is nothing but an advancing alternating electromagnetic field. If this grazes the atom, it gives rise to a dipole which vibrates in unison with the light-field. This dipole then emits the spherical scattered waves described previously. " Oh yes, then the air, and indeed every body, must always scatter light—but we said before that scattering is associated with variations of density of the

atoms?"—Both statements are true, but we must
remember the existence of interference. Each
separate atom scatters light; but if the atoms are
distributed with perfect uniformity, the scattered
waves interfere and cancel one another. This can
be explained by an argument similar to that used to
explain the sharp boundaries of shadows (p. 112):
if a hole in a screen is many times greater than the
wave-length, the secondary waves produced at the
hole by the wave-crests and wave-troughs annihilate
one another in the region of shadow, and reinforce
one another elsewhere. In exactly the same way, the
scattering of light by atoms distributed uniformly
in space only results in a trifling change in the light
passing through (namely, in its velocity and intensity), but gives no appreciable intensity of light in a
lateral direction. If, on the other hand, the distribution of the particles is irregular, the light scattered
laterally is not completely destroyed, and the blue
of the sky results.

The point we are concerned with here is this.
According to the wave theory, the wave-length of
the scattered light must be exactly the same as that
of the incident light. Frequency and wave-length
must be unaltered by scattering.

In reality, however, there are cases where this
does not happen, namely, with light of very short
wave-length (X-rays). The discrepancies are only
small; but it is the distinguishing mark of a good
scientist that he pays attention to small discrepancies,
observes them accurately, magnifies them and elu-
cidates them until he has clearly worked out a defect

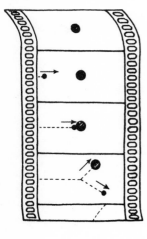

(58)

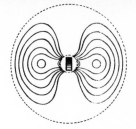

in the accepted theory. Then the way to a new discovery lies open.

Compton investigated the scattering of X-rays by paraffin wax, and found that the rays scattered sideways or backwards have a wave-length slightly greater than that of the original radiation (Plate IV(*a*), facing p. 254). On the wave theory this is unbelievable. For at bottom it should be just a matter of a mechanical process resembling the steady shaking of a plum tree, when of course all the twigs shake in the same rhythm.

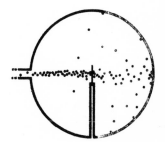

Compton recognized, however, that the phenomenon is immediately intelligible on the photon hypothesis.

Paraffin wax is a substance consisting entirely of light atoms, carbon and a great ·deal of hydrogen. In these light atoms the electrons are held comparatively loosely, so that the atoms are easily ionized. In addition, an X-ray photon is very powerful; its energy is *h* times the frequency, and the frequency is perhaps 10,000 times that of visible light. With an impact of this magnitude the effect of the binding of the electrons is altogether negligible. In this process the block of paraffin wax may be regarded as a collection of free electrons.

Now we simply have a kind of billiards with two different balls. One ball is an electron in the paraffin wax; it is almost at rest. The other ball is the photon rushing at it (58).

If a collision takes place, the balls fly apart, according to the laws of mechanics; the electron receives kinetic energy. As the total amount of energy must remain the same, the rebounding photon must have less energy than before. By Planck's law, this means that it has a lower frequency, that is, a longer wave-length.

What we have pictured qualitatively here is not difficult to work out quantitatively, even when the relativistic change of mass (p. 82) is taken into account.

Observation has completely confirmed the laws obtained in this way. In particular, the recoil electrons can actually be observed. If the paraffin wax is replaced by a gas, the tracks of the recoil electrons can be made directly visible by means of the Wilson chamber. The photon cannot be seen flying away, but often it liberates an electron photo-electrically from the wall of the chamber. This is visible as a cloud track and its direction is that of the photon. The latter must bear a definite relation to the direction of the recoil electron. This, too, is verified by experiment. Further, electrons and photons (or rather, the secondary electrons produced by the latter) can be caught in separate counting arrangements, and it is found that the two counters are affected exactly at the same instant. Finally, the absolute magnitude of the change of wave-length agrees with the calculated value. The latter is determined by a single number, namely, the change in wave-length when the photon is deflected at right angles to its original direction. Theory gives this value as $\frac{h}{mc}$, where m is the mass of the electron. This is a length, which is called the Compton wave-

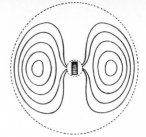

length; when the numerical values are inserted it is found to be $2 \cdot 42 \times 10^{-10}$ cm. This, too, is in good agreement with the experimental results.

There is thus no doubt that the photons behave like billiard balls when they collide. On the other hand, there is just as strong evidence that light behaves like waves.

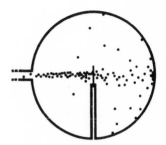

This is a horribly awkward situation for a science which rather prides itself on its rational modes of thought.

At first scientists met the difficulty in a purely intuitive way. They found that in certain phenomena the wave ideas were appropriate, in others the photon ideas. As Sir William Bragg has said, " On Mondays, Wednesdays, and Fridays we adopt the one hypothesis, on Tuesdays, Thursdays, and Saturdays the other ". The first group of phenomena depended on the distribution of light intensity in space, like interference bands, the resulting spectra, and so on; the second group, on the transformation of light energy into other forms of energy, or conversely. But there are some experiments in which both ideas are found necessary at the same time. For example, instead of looking at a system of interference bands with the eye, it is possible to make a small counting apparatus pass over them and count the electrons photo-electrically liberated by them in the tube (from the walls or the gas within it). Then we obtain many electron impacts just when the counting tube is over a bright band, and few when it is over a dark· band.

There is no difficulty in uniting the two aspects

if the illumination of a screen showing interference bands is so strong that the number of photons falling on unit area of the screen during the time of the observation is very large. Then it is clear that this number, multiplied by the energy ($h\nu$) of one photon, must give exactly the intensity of the light deduced by calculation from the wave theory. Where this intensity is strong, that is, in the bright interference bands, many photons arrive, whereas in the dark bands, no photons, or only very few, arrive. The illumination as determined by the waves regulates the supply of photons, but the dynamical action of the light, in liberating electrons, say, is given by the energy of the single corpuscle, the photon.

So far everything is simple. The intrinsic difficulty of reconciling the wave aspect and the corpuscular aspect only arises when we have to deal with very feeble illumination, where the total energy of the light falling on the screen during the time of observation is not much greater than the energy of a single photon.

The photon must arrive somewhere on the screen; but where? Has the light-wave some properties, as yet undiscovered, which regulate the path of the photon by causal laws? Certainly we know of no such property. And observation shows that the electrons struck out from a metal surface by feeble light appear at random; their distribution in space and time follows the laws of probability, with the sole restriction that for long runs, that is, for strong light or long periods of observation, the mean dis-

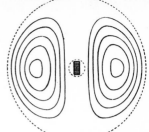

tribution exactly follows the law predicted by the wave theory.

This fact can be expressed by the statement that the intensity of the waves at a given place determines the probability of finding a photon there. For the expression " probability of an event " simply means the number of cases where this event happens, averaged for a very long run.

It has been said that the electromagnetic waves have consequently been degraded to play the part of " waves of ignorance ". I should say the opposite is true. They give everything which can be said with certainty about the appearance of photons, namely, the average number of these particles; it would be better to call them " waves of partial knowledge ". Here the word " partial " is important. For our line of thought suggests that the wave aspect and the corpuscular aspect cannot be unified without our giving up the idea that everything can be predicted from causal laws. We may suspect that the fundamental ideas of physics on " cause and determinism " need a very thorough revision.—But first a still greater catastrophe was to come.

7. *Electron-waves.*

Many people say that the only use of a theory is to spur on the investigator to fresh researches. I am not of this opinion. As I have already emphasized repeatedly, no experiment has any *meaning* at all unless it is interpreted by the theory. I do not say *a* theory, but *the* theory. For it is only to onlookers that contradictory theories appear to contend with

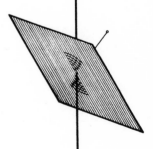

one another and to have their spells of authority.
Granted, this contradiction often appears to occur;
but this is only due to the fact that, at the moment,
the facts are not sufficient to permit of a definite
decision.   So long as this is so, both theories are
really equally justifiable, and so far as the realm of
observable phenomena is concerned they are just
different forms of the same theory.   The mathematician says that one can be " transformed " into
the other.

As an example we may take the dispute over the
question whether electrical forces act at a distance
or through a medium. At the beginning of the nine-
teenth century most physicists, particularly on the
Continent, asserted that the electric force acts
through the empty space between two charges
(" action at a distance ").   Faraday, however, developed the idea that the electric field is due to
something which happens in the space between the
charges (" action through a medium ").   The controversy over this question was a bitter one, as
always happens with matters which cannot be settled
once and for all.   For the mathematicians proved
beyond any doubt that both representations are
exactly equivalent and must always give the same
results—so long as experiments were not sufficiently
accurate to decide whether electrical disturbances
were propagated instantaneously or required a finite
time. As soon as Hertz had shown by the discovery
of his waves that the velocity of propagation is finite,
the quarrel died down of itself.

We are therefore convinced that there is only *one*

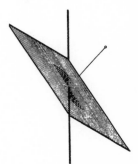

theory, to which we gradually approach. At every
stage of approximation there are several possibilities
of advance, round which disputes rage until a new
discovery settles the matter one way or the other.

It often happens that a theory has astonishing results, namely, it can predict quite new phenomena,
of which the experimenter would never have thought.
The theoretical physicist must also consider the
possibility of improving and refining the still incomplete and ambiguous theory. This "refinement"
is a curious thing: in actual fact, a sort of æsthetic
feeling plays no small part in the deliberations of
the theoretical physicist. Thus, for anyone who has
a command of mathematics, Einstein's theory of
relativity seems more complete and satisfying, in-
tellectually, and so more beautiful, than Newton's
mechanics. I believe that this feeling really arises from
the removal of an arbitrariness and ambiguity in the
older ideas, which was felt as a disturbing influence.

One of the most astounding prophecies was made
by de Broglie in 1925. His train of thought was this:

In optics there is the antithesis of wave and
photon. That a hundred years ago people believed
in waves only, was because the phenomena of interference are comparatively simple to observe. The
photo-electric effect, on the other hand, required
the whole resources of modern electrical engineer-
ing, and was therefore long in being discovered.

In the case of the cathode rays and other electrical
rays the experiments forcibly suggested the corpuscular idea from the very beginning. Is it necessarily true that here the wave idea is quite off

the track? Should not the antithesis of wave and particle apply to the electron also?

De Broglie proceeded to deduce the consequences of this bold idea, using all the facts which were known in the case of light.

In the first place, he had to take account of the theory of relativity. According to it, lengths and intervals of time have to be modified in certain ways if we pass from one frame of reference to another (moving uniformly). In this transformation, lengths and intervals of time occur symmetrically. Now the period of a vibration of light (τ) is an interval of time; it is the reciprocal of the number of vibrations per second, and Planck's law can be expressed in the form

$$E\tau = h.$$

By the theory of relativity, this interval of time, the " period " of the light vibration, must have an analogous spatial quantity corresponding to it. What can this be? Obviously the spatial period of the waves, i.e. the wave-length. What corresponds to the energy, then? We again turn to the theory of relativity for guidance and learn that the energy (E) and the momentum (p) of a particle go together; in the passage from one frame of reference to another they are transformed in the same way as intervals of time and lengths. We are therefore led to suppose that in addition to Planck's law there is another law, de Broglie's law, namely,

$$\text{Momentum} \times \text{Wave-length} = h,$$

or $$p\lambda = h.$$

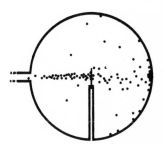

But in the case of light this is nothing new. For we
have seen that the momentum of a light-wave is
equal to its energy divided by the velocity of light
$\left(p = \dfrac{E}{c}\right)$; on the other hand, the wave-length is
equal to the period multiplied by the velocity of
light ($\lambda = \tau c$). In the product $p\lambda$ the velocity of
light cancels out and we are left with Planck's law.

It is otherwise when we are dealing, not with
light, but with electrons. In the first place, their
velocity is only a fraction of the velocity of light.
In the second place, the energy and the momentum
do not depend in such a simple way on the velocity.
It follows that here de Broglie's law will tell us
something new.

We now imagine that the electron has a wave
accompanying it, the *de Broglie wave*, which bears
the same relation—still an obscure one—to the
electron as the light-wave bears to the photon.

For any wave motion whatever, the velocity of
propagation (u) is given by

$$\nu\lambda = \frac{\nu}{\kappa} = u,$$

where ν is the frequency, λ the wave-length, and κ
the wave-number, i.e. $1/\lambda$, or the number of waves
per unit length.

We can multiply the numerator and denominator
by h, and we then have

$$u = \frac{h\nu}{h\kappa} = \frac{E}{p},$$

by the formulæ of Planck and de Broglie. On the other hand, the theory of relativity enables us to express E/p in terms of the velocity of the particles; for E is mc^2, and p is simply mv, so that E/p is c^2/v, giving

$$u = \frac{c^2}{v} \quad \text{or} \quad uv = c^2.$$

In this way we get u, the velocity of the de Broglie wave, in terms of v, the velocity of the particles. For light the denominator is equal to one of the factors of the numerator, so that in this case the velocity of propagation again comes out as the velocity of light.

In the case of electrons, however, the velocity of the particles is always less than the velocity of light. Hence it follows that the de Broglie guiding wave ("pilot wave") travels faster than light. This seems an unhappy result; for the whole theory of relativity depends on there being no measurable velocity greater than that of light. In reality, however, this is the clue to the right understanding of the remarkable dual entity of wave and corpuscle. To see this we must consider the idea of a wave somewhat more closely.

8. *Wave Groups and Group Velocity.*

Hitherto we have talked in a very vague way about waves, meaning any vibratory motion which is propagated. We must now express ourselves more accurately.

If we ascribe a definite time-period and a definite

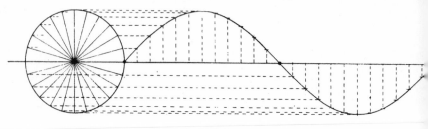

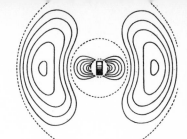

wave-length to a wave, we are really referring to a special kind of wave, which we should more accurately describe as a *simple harmonic wave*—and which does not really exist in Nature at all. We can represent this wave on a piece of paper, taking the direction of propagation to the right and the wave amplitudes upwards (59).

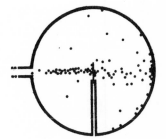

The wavy curve thus obtained is unbounded in both directions, and is completely determined by its wave-length and amplitude. If we think of this wavy curve as advancing uniformly, say to the right, we have a progressive simple harmonic wave, which goes on for all time, with a definite velocity of propagation and a definite period, at every fixed point.

This indefinite extension in space and time is really part of the concept of a simple harmonic wave. Only then is there any sense in distinguishing the wave by giving its wave-length and period (as well as the amplitude).

Every other kind of wave-motion, such as one which starts at a certain point and stops at a certain point, requires many more data to describe it—the first and last points at least.

When we have spoken of waves hitherto, we have always tacitly meant these simple harmonic waves. It is, however, unfortunate that such do not exist in Nature—or only approximately; in the case of light, for example, we have wave-trains extending to two or three millions of simple harmonic periods.

Simple harmonic waves, then, are just an ideal limiting case, a useful concept for analysing more complicated wave effects.

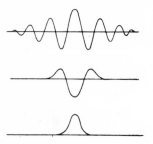

Simple harmonic waves cannot carry a signal. They are smooth as oil and each wave-crest is exactly like every other, and they go on monotonously for ever. Hence they may very well travel faster than light, for the theory of relativity would only be contradicted *if they could be used to send distinguishable signals for the comparison of clocks.*

As we have seen, de Broglie's wave has a velocity greater than that of light; we have ascribed to it definite space and time periods, that is, we have regarded it as a simple harmonic wave. Then the contradiction to the theory of relativity disappears. A new difficulty arises: if the association of particle and wave is to be possible, the particle must move uniformly in a straight line. But it certainly started its motion at some point and it will certainly end it at another point. Thus we are forced to ascribe a beginning and an end to the pilot wave. Hence it cannot be a simple harmonic wave. It is certain, therefore, that the association of particle and wave is nothing but an approximation which corresponds to an ideal case. We must try to get a little closer to the reality.

To do this we must consider how a number of simple harmonic waves are combined and what the result is. In the first place we consider a wave at a definite instant; in a sense we think of a snapshot of the wave. The amplitudes then form a certain curve, like the three curves shown in figure (60). All three begin at a certain point and end at another point. The uppermost curve still exhibits a definite wave-like character; a period can be recognized. In

(60)

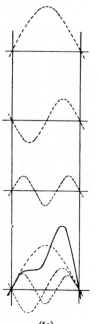

(61)

(62)

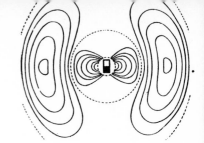

the second, very little trace of this is left, and the third is just a kind of hump, with no sign of periodic repetition.

Now we take a crowd of simple harmonic curves differing in wave-length and amplitude and superpose them, that is, we simply add their amplitudes. We then obtain all manner of curves—of which we give an example opposite (61).

About a hundred years ago the mathematician Fourier excited great attention and loud controversy by his assertion that, by suitable choice of a great many simple harmonic curves, any prescribed curve whatever can be represented as closely as desired as the sum of simple harmonic curves.

This seemed perfectly unbelievable. What about curves with sharp corners, like the one in (62)?

Yet Fourier was able to prove with mathematical strictness that his assertion holds good in every case. True, to represent sharp corners decently, we must have a very great number of superposed simple harmonic curves; but figure (64), p. 148, shows that with not too large a number of curves it is possible to obtain at least a similar curve with rounded corners. One thing is clear: the more a curve deviates from the shape of a simple harmonic curve, the greater the number of simple harmonic curves which must be combined in order to represent it. There is a simple theorem about this, which enables us to estimate, at least roughly, how many simple harmonic curves are required. It is as follows:

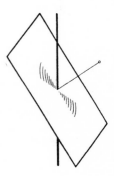

A curve which exists only in a definite interval of space, and outside this interval is zero, does not

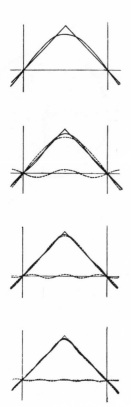

(64)

require curves of all possible wave-lengths for its representation, but chiefly those whose wave-numbers (numbers of waves per unit length) lie in a definite region. Here the word " chiefly " means that the waves which do not fall into this region are very feeble. We shall ignore them. Further, we have the qualitative statement that the product (Extension of the Curve in Space × Interval of Wave-numbers) is approximately equal to 1, or, in other words, these two quantities are the reciprocals of each other.

This theorem is an abstract expression of the physical phenomenon of diffraction at a slit. We have seen (p. 110) that the breadth of the slit and the breadth of the diffraction pattern on the screen are inversely proportional to one another (63). If the intensity of the light passing through the slit is represented by a curve, it is of exactly the same type as that described above. The diffraction pattern, however, is the decomposition into simple harmonic waves; for it arises from the superposition of the elementary spherical waves coming from the different points of the slit. This mathematical law therefore immediately explains why a narrow slit gives a broad diffraction pattern.

Instead of trains of waves at rest, we shall now consider progressive waves. So long as they all have the same velocity, the figure obtained by their superposition just moves forward as a whole. This is true of light. A ." packet " of plane simple harmonic waves moves on through space un-distorted.

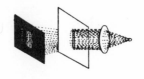

(63)

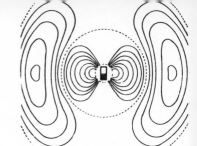

With electron-waves it is not so. Here simple harmonic waves of different wave-lengths have different velocities. Owing to this, the figure obtained by superposition is distorted in an extremely complicated way. And yet a simple law holds good even here.

We shall consider the case where there are only two waves of almost equal wave-length and hence of almost equal velocity. This case is shown in Film IV, which illustrates the two simple harmonic waves and the wave obtained by their superposition. Here we have the phenomenon of *beats*, with which everyone is probably familiar in the acoustic case. If, for example, a piano is badly tuned, so that the two or three strings belonging to the same note are not vibrating at exactly the same rate, the tone is heard to quaver, increasing and diminishing in loudness. The same is shown in the " film ". We see that the wave resulting from the superposition has periodic fluctuations of amplitude; to show this clearly, the maxima are joined by a curve, which again has the form of a simple harmonic wave, of long wave-length. If we were dealing with light (or sound), where both of the primary waves advance at the same rate, this beats-wave would move at the same rate and produce the fluctuations of intensity. On looking at our moving pictures, however, we notice at once that this is not so: the primary simple harmonic waves move with slightly different speeds, the longer wave more rapidly than the shorter, and as a result the beats-wave moves much more slowly than either.

Exactly the same thing happens if we have, not two primary waves, but a whole " group " of waves, a " wave-packet ", the wave-lengths of which are all confined to a narrow interval. Then by Fourier's theorem (p. 147), the whole wave-packet is fairly extended in space, just as the single beats " hump " in the case of two primary waves is long compared with their wave-lengths. The velocity of the whole packet, the so-called *group velocity*, however, is much smaller than the velocity of the individual waves, provided that this increases with the wave-length. To distinguish them, the latter velocity is called the *phase velocity*.

De Broglie made use of these facts to give a more accurate description of the relation between the motion of the electron and the pilot wave. He actually worked out the group velocity and found that it is less than that of light and exactly equal to the velocity of the electron—a result as astonishing as it is gratifying. The simple harmonic waves are obviously a mathematical fiction—they extend indefinitely and move with a velocity greater than that of light. Actually the uniformly moving electron has associated with it a group of waves with an extremely narrow range of wave-lengths (and hence a great extension in space), and this group keeps step with the electron, moving along with it with the same velocity. But what is the meaning of the extension of the wave-group in space? Before we answer this question—we have, I think, indulged in plenty of theoretical speculation about it—we prefer to find out first whether cases actually occur in

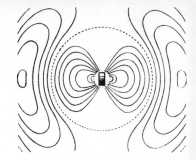

Nature where the electron-waves can be detected experimentally.

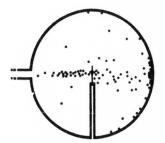

9. *The Experimental Detection of Waves of Matter.*

The electron-waves were not at first systematically sought for, but were discovered as a by-product of experiments with quite a different aim. Davisson and Germer were occupied with investigations on the reflection of electron beams by metals. Quite remarkable irregularities in the intensity of reflection showed themselves. Elsasser first hit on the idea that there might be a connexion between these observations and de Broglie's hypothesis. The lattice or grating of metal atoms forms the connecting link. For we have seen already that the regular configurations of atoms which we call crystals act like artificial gratings, giving rise to interference patterns when light passes through them. In general, however, pieces of metal are not themselves crystals, but consist of a confused mass of very minute crystals. Hence the interference patterns produced when X-rays pass through a layer of metal are not simple bands or spots, but rings; a single little crystal produces an interference spot in a definite position relative to the incident ray, and the spots arising from all the little crystals coalesce into a bright circular ring.

Plate III(*b*), facing p. 250, shows one of these photographs for X-rays, and Plate III(*c*) a corresponding photograph for electron waves. This was not taken by the Davisson-Germer reflection method, but in exactly the same way as the X-ray photograph,

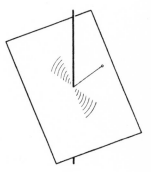

by making cathode rays pass through thin metal foil and fall on a photographic plate. We see that the two pictures are very much alike and immediately suggest an analogy between the two phenomena. If one arises from interference, so must the other. But, if there is interference, the cathode rays must have a wave character! More than this, de Broglie's law, which may be stated in the form

$$\text{Wave-length} = \frac{\text{Planck's Constant}}{\text{Momentum of Electron}},$$

is actually confirmed quantitatively. For it is very easy to accelerate or retard electrons, that is, to alter their momentum, by making them pass through an electric field, say between the cathode and a grid. If the momentum is doubled, the wave-length must fall to half and the interference rings must likewise shrink to half their size. This is actually what happens. From the size of the rings we can conclude that the wave-length of the electrons is of the same order of magnitude as that of X-rays. As we can also measure the velocity of the electrons, we have all the data for testing whether Planck's constant is really obtained from the formula, and the result comes out right.

Looking back over the development of electron theory, we must wonder why electron interference was not discovered earlier; for any number of physicists had made observations on the passage of electrons through sheets of metal! Really it is a piece of good luck that they overlooked the rings. For otherwise what confusion would have arisen!

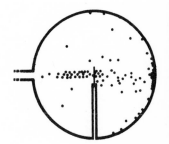

Two parties would have arisen, the " corpuscle "
party and the " wave " party. They would have
disputed with all the heat and passion which always
come to the surface when genuine arguments are
wanting. And at that time—say thirty years ago—
conditions were not ripe for the recognition that
perhaps both parties were in the right. Within the
science which prides itself on being the most exact
of sciences, we should have had a state of affairs
known only in connexion with religious, philosophical, or political questions. " Particle-man "
would have been a term of abuse among wave-
fanatics, just as the word " Bolshevik " is among
Fascists and vice versa. By good fortune the history
of the human mind has been spared this disgrace.

Does the dual entity of wave and particle exist for
electrons only? For there are other forms of material
rays—beams of ions of every kind, and even of
neutral atoms or molecules, such as we mentioned
in Chapter I (p. 28). Should not these genuine
material rays also give interference patterns under
suitable conditions?

This does really happen. For example, Stern has
produced molecular beams of hydrogen and helium
and let them be reflected at a small angle from the
surface of a crystal. Then the ordered ions of atoms
in the crystal act like the rulings of an optical
grating, and we obtain a genuine spectrum of
molecules, just as if they were photons. For of
course the beam of molecules contains particles of
different velocities, that is, different de Broglie wave-
lengths, just like white light. We can go even

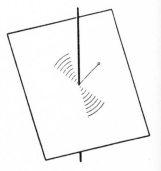

farther than this. By mechanical apparatus we can select from the beam a narrow bundle of particles, all of which have almost the same velocity. An example of such apparatus is the rotating apparatus shown in Film II; in another, two toothed wheels rotating on the same shaft are used, as described on p. 30; this is more convenient for selecting particles of a given velocity. These particles are then made to fall on the surface of the crystal, are reflected there, and are finally caught in a receiver.

The result completely confirms de Broglie's ideas. Even ordinary atoms and molecules, when in the form of beams, behave, as regards their *distribution in space*, according to the laws of the wave theory.

It is equally clear that as regards their mechanical effects they retain the character of particles throughout. For the receiver might take the form of a counting apparatus, and we should then find that at the interference maxima many particles are collected, at the interference minima but few.

The discovery of electron-waves and material waves is perhaps the most important achievement of physics in our time. As a result, essential alterations have been made in the fundamental ideas of the science. These we must now discuss.

10. *Wave Mechanics and its Statistical Interpretation.*

Experiments demonstrate quite clearly that light and matter unite in themselves properties of waves and properties of particles. We therefore cannot say that they are one or the other: they are both, displaying one side of their nature or the other,

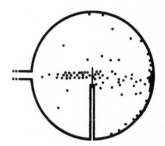

according to the method of investigation.

This circumstance raises great difficulties of theoretical interpretation. Bohr has declared outright that there is an incomprehensible irrational factor in physical events. To make the position clear, we need only state bluntly what the quantum postulate of Planck and de Broglie means.

Energy and momentum are properties of minute particles. Frequency and wave-number, on the other hand, are properties of simple harmonic waves, whose definition implies that they extend indefinitely in time and space.

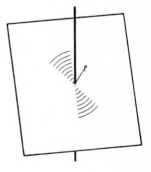

Yet it is asserted that—apart from the factor h, which serves to transform the units of measurement —energy and frequency are to be identified, and also momentum and wave-number.

We see at once that this is not possible unless we sacrifice some fundamental assumption of ordinary thought.

The case is like that of the theory of relativity. There experiments on the behaviour of light in rapidly moving systems forced us to form a new conception of space and time. Here in the quantum theory it is the *principle of causality*, or more accurately that of *determinism*, which must be dropped and replaced by something else.

To be quite clear about what this principle means, we shall return to the illustration of a gun firing, which we used right at the beginning of the book. We then said (p. 17) that a knowledge of the laws of nature is far from being sufficient to enable us to make predictions about future events; we must

know the initial conditions as well. In the case of
the gun the form of all possible trajectories is determined by a law of nature, which expresses the effect
of gravity (and perhaps also air resistance) on the
motion of the shell; but the path actually followed
by the shell in a prescribed case depends on the
direction in which the gun is trained and the muzzle
velocity of the shell.

Now in the older physics it was assumed as
obvious that these initial conditions can always be
stated with any desired degree of accuracy. Then
the course of the subsequent phenomena (the tra-
jectory of the shell, in the case of the gun) can also
be calculated with any desired degree of accuracy.
The initial state determines the future according to
the laws of nature. From a given state onward
everything goes on like an automatic machine and,
provided we know the laws of nature and the initial
state, we can predict the future merely by processes
of thought and calculation.

This actually does happen. Astronomers, above
all, predict the positions of the moon and the
planets, the occurrence of eclipses, and other celestial
phenomena, with great accuracy. Engineers, too,
rely firmly on their machines and structures doing
what they have been calculated to do—and successfully.

Nevertheless, modern physics declares that the
matter is not so simple as this, whenever we have to
deal with the restless universe of atoms and electrons.

Even in the case of gases we saw that the determination of phenomena from the initial state may

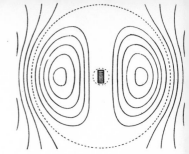

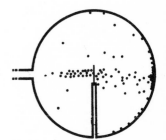

be an excellent idea in theory, but is of no practical consequence. For it is quite impossible to determine the positions and velocities of all the particles at one instant. Instead, we have recourse to statistics. We make an assumption about equally probable cases (the hypothesis that the molecules are arranged at random) and deduce results from this. As these agree with experiment, we are led to the belief that statements about probabilities can be just as good objective laws of nature as the ordinary laws of physics. This kind of statistical argument, however, has only a loose and superficial connexion with the rest of physics.

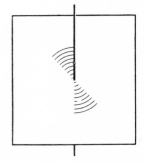

All this is changed by the discovery of the dual entity of wave and particle. Experiments show that the waves have objective reality just as much as the particles—the interference maxima of the waves can be photographed just as well as the cloud-tracks of the particles. There seems to be only one possible way out of the dilemma; a way I have proposed, which is now generally accepted, namely, the *statistical interpretation* of wave mechanics. Briefly it is this: *the waves are waves of probability*. They determine the " supply " of the particles, that is, their distribution in space and time. It follows that the waves, apart from their objective reality, must have something to do with the subjective act of observation.

Here lies the root of the whole matter.

In the older physics it was assumed that the universe goes on like a machine, independently of whether there is someone observing it or not. For the observer was believed to be capable of making his observations without disturbing the course of events. At all events, an astronomer looking through his telescope does not disturb the march of the planets!

But the position of a physicist who wishes to observe an electron in its path is not so simple. He is like a craftsman who is trying to set a valuable diamond with a mason's trowel. He has no apparatus available which is smaller and finer than the electron. He can only use other electrons, or photons; but these have an intense effect on the particles under observation, and spoil the experiment. We see that a necessary consequence of atomic physics is that we must abandon the idea that it is possible to observe the course of events in the universe without disturbing it.

Now if the steps necessary for making an observation had quite complicated effects on the events, mathematical physics could not exist at all. Happily this is not so. The fundamental laws of the quantum theory, with which we are already acquainted, see to it that enough is left to enable us to make predictions. But the predictions are no longer " deterministic ", in the sense that " the particle observed here to-day will be at such and such a place to-morrow "; but " statistical ": " the probability that the particle will be at such and such a point to-morrow is so and so ". In the limiting case of large masses, such as we have in ordinary life, this probability of course becomes practical certainty; here the principle of causality still holds in its old form.

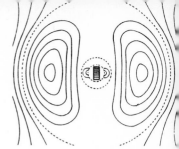

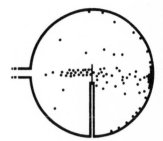

To penetrate more deeply into the meaning of these statements, we go back and consider an electron and its pilot wave. We saw that physically there is no meaning in regarding this wave as a simple harmonic wave of indefinite extent; we must, on the contrary, regard it as a wave-packet consisting of a small group of indefinitely close wave-numbers, that is, of great extent in space. Then the group velocity is identical with the velocity of the particle; the wave-packet moves with the particle. But whereabouts in the packet is the particle?

Clearly it is in accordance with the spirit of the probability idea to say that this question has no answer. We can, however, say that the particle has an equal probability of being anywhere in the wave-packet. The wave is just that part of the description of the phenomenon that depends on the intrusion of an observer; it replaces the initial conditions of classical physics. The difference, however, is this: the assumption that the particle has a definite velocity necessarily means that the position of the particle is and must remain largely indeterminate. For it is only in the case of a group of waves of almost equal wave-numbers that we can speak of a group velocity. As was stated on p. 148, the product

Extension in Space of Wave × Range of Wave-number

is approximately 1; hence if the range of wave-number is small, the extension in space must be great.

This rule can be stated in another way, in which it no longer refers to waves but tells us something

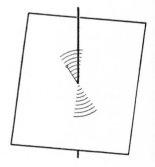

about the measurability of the position and velocity of the particle.

As we have seen, the extension of the wave in space corresponds to the uncertainty about the position of the particle. We now recall de Broglie's relation:

$$\text{Wave-number} = \text{Momentum} \div h.$$

A definite range of wave-number therefore corresponds to a definite uncertainty about the momentum. Thus we obtain the result that the product

$$\text{Uncertainty of Position} \times \text{Uncertainty of Momentum}$$

is never less than h. This is the celebrated *Uncertainty Principle* of Heisenberg, which interprets the irrationality of the quantum laws as a limitation of the accuracy with which various quantities can be measured. There is another similar relation between time and energy. We shall now illustrate the meaning of these relations by examples.

On the particle theory the phenomenon of diffraction at a slit is to be interpreted as follows:

The slit gives the indeterminacy in the position of the photon in a direction at right angles to the direction of the ray. As we know, the narrower the slit, the broader the diffraction pattern on the screen. According to the uncertainty principle, in fact, the momentum of the particles in the plane of the slit is more indeterminate, the more accurate the determination of position in that plane is, that is, the narrower the slit is. This means, however, that the particles spread sideways.

A second example is the observation of an electron

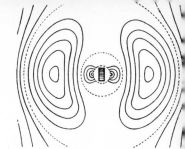

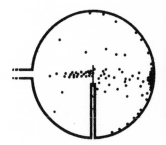

through a microscope. It is obvious that under the microscope we can never distinguish distances which are smaller than the wave-length of the light used. To determine the position of the electron accurately, we must accordingly use light of short wave-length. But then the Compton effect (p. 133) appears: the light exhibits its corpuscular properties and gives the electron a push in a direction which lies within the aperture of the instrument. In the determination of position, the electron has therefore acquired an uncertainty of momentum, and a simple argument shows that Heisenberg's relation again applies. If we try to avoid the Compton effect by taking light of longer wave-length, the uncertainty in the momentum becomes less, it is true, but the uncertainty in the position is increased.

A great number of similar examples may be adduced. The further development of the quantum theory has shown that there are other pairs of quantities, besides position and momentum, which cannot be measured simultaneously to any desired degree of accuracy. Increasing the accuracy with which one quantity is measured decreases the accuracy with which the other can be measured. Such quantities are said to be *conjugate*.

From the corpuscular point of view, therefore, Planck's constant h means that there is a *natural limit* to the accuracy with which conjugate quantities can be measured.

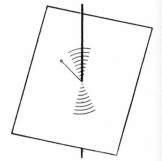

By this renunciation of the ideal of unlimited accuracy of observation, the irrationality of the quantum postulate is removed in this sense, that no contradictions can arise from any actual experiment. One has only to give up the idea that there exist ideal experimental arrangements with the help of which any question of interest about a physical system could be decided. In fact there are different experimental arrangements which are mutually exclusive but complementary, and only by performing all of them can one obtain the maximum information possible.

This idea of "complementarity" was introduced by Niels Bohr to describe with a short expression the new standpoint in regard to the relation of experiment and theory. He has given many instructive examples for complementary situations. Consider, for instance, Young's interference experiment with the two slits (p. 110). The interference pattern represents the probability of finding a photon on a given spot of the screen; in fact, by using a counter instead of a photographic plate one can register the arrival of each single photon and thus confirm the particle nature of light and statistical interpretation of the waves. Hence it seems to be legitimate to ask: through which of the two slits has a certain photon passed? In fact, whether this question is legitimate or not depends on the meaning of the word "certain" in the last sentence. If it means just "some" photon it is possible to answer it; one has only to attach to one of the slits an indicator reacting to the passage of the photon (for instance, one can arrange the slit itself on a light mobile frame which can receive momentum from the photon). The passing of the photon is observable by the reaction of the indicator

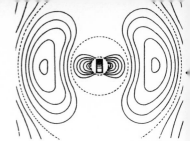

(recoil of the frame, say). But then the photon will suffer an unpredictable deflection which is only statistically determined, hence it will reach the screen at some other place than before, and the interference pattern will be spoiled. In fact this pattern depends on the assumption of a rigid connection of all parts of the instrument. Therefore, if "certain" means a photon reaching a fixed place on the screen and being actually observed there the mechanism at the slit must be put out of action.

Other examples of such inconsistency of two physical questions and of the complementarity of the corresponding apparatus has been discussed by Bohr. In every case it could be shown that no real, that is experimentally demonstrable, contradictions can arise.

We now come back once more to the question of causality. This rests on the assumption that the initial state (that is, the initial position and the initial velocity or momentum) can be determined exactly. It is found, however, that actually only one of the two can be determined with perfect accuracy (or both inexactly). Hence the law of causality becomes meaningless. Many cling fast to it "because it is a necessity of thought". Galileo's opponents declared that it was a " necessity of thought " that the earth is at rest at the centre of the universe: how else could it be the home of man, the crowning glory of creation? " Necessities of thought " are often just *habits* of thought.

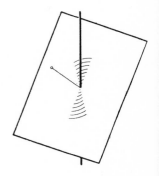

We now have a *new form* of the law of causality, which has the advantage of explaining the objective validity of statistical laws. It is as follows: if in a certain process the initial conditions are determined as accurately as the uncertainty relations permit, then the probabilities of all possible subsequent states are governed by exact laws. If the experiment is repeated a great many times with the same initial conditions, the frequency of occurrence of the expected effects (and its fluctuations) can be predicted.

In the case of the large bodies of everyday experience we come back to the old law of causality. For, as the momentum is equal to Mass × Velocity, in the case of a large mass a definite fluctuation of the momentum will only give rise to a very small fluctuation of the velocity. Position and velocity can therefore be determined simultaneously with sufficient accuracy.

Even in the case of atoms or ions, which are many thousand times heavier than electrons, ordinary mechanics often applies with a very high degree of accuracy.

People who judge the value of discoveries by their influence on practical life need not be troubled about these new conceptions. They can leave them to the specialists who have to deal with electrons and photons in physical laboratories. Those who are interested in the philosophy of science, however, cannot pass over the fundamental ideas which have arisen out of physics, but may have much wider consequences. Bohr has pointed out that there may be a similar kind of complementary relation between the biological aspect and the psychological aspect of the processes of mind. It seems impossible to investigate the state of brain connected with some idea

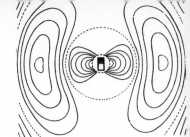

or sensation by biological methods without essentially altering the psychological content of this idea or sensation. You cannot experiment on the physiology of the brain without using electric currents, chemical reactions, or even surgical operations, which affect the sensations of the victim. It may be that this is not merely a question of technique, but that there is a law of nature which prohibits physical and chemical statements on nervous processes being made with any desired degree of accuracy simultaneously with statements on their images in consciousness—just as the quantum principles prohibit the simultaneous measurement of position and of velocity. Then the old philosophical question of the parallelism between the physical phenomena in the brain and the correlated psychical phenomena would become meaningless and would have to be formulated in a new way.

This rather difficult chapter was intended to enable us to explain the structure of atoms. It has, however, led us to deeper and more general considerations.

The completest knowledge of the laws of nature does not carry with it the power of prediction, nor of mastery over Nature. *If* the universe is a machine, its levers and wheels are too fine for our hands to manipulate. We can learn and guide its large-scale motions only. Beneath our veiled sight it quivers in eternal unrest.

" And the Spirit of God moved upon the face of the waters."

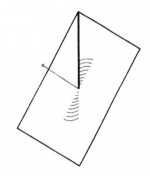

CHAPTER IV

The Electronic Structure of the Atom

1. The Discovery of the Positive Charges in the Atom.

NOW we know enough about the electron to be able to study the structure of the atom accurately.

The first question is one which we have already asked more than once, but have not yet answered: where is the positive electricity in the atom?

For, as the whole atom is electrically neutral and it has negative electrons in its outer regions, there must be a positive charge somewhere inside.

We must therefore penetrate into the interior of the atom. For this purpose we shall find fast particles useful, as they have great momentum. The first systematic investigations of this kind were made by Lenard, using fast cathode rays. As we have already mentioned (p. 94), these will pass through thin layers of metal. In solid bodies the atoms are tightly packed together. The outer shells of electrons of one atom are almost touching those of the next. If an electron flies through with only a small deviation of direction it must have passed through the interior of the atom almost undisturbed (65). This

ELECTRON

+METAL+

(65)

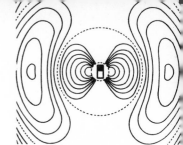

interior must therefore be comparatively empty, in spite of its compact outer layer.

Occasionally, an electron is strongly deviated; we naturally suppose that it has struck some solid obstacle. Lenard was able to explain the bulk of his observations by assuming that in the interior of the atom there are a number of heavy positively charged particles, which he called " dynamids ".

We shall not, however, use these experiments to explain how such conclusions can be drawn, but those of Rutherford instead. Rutherford used heavier projectiles, the α-particles from radioactive substances, which we have already mentioned (p. 102). The advantage of these is that they are not deviated in the slightest by collision with the much lighter electrons; they register only their collisions with particles which are at least about as heavy as they are themselves.

The experiments showed in the first place that most of the α-particles pass through thin metal foil without deviation, but a very minute fraction are noticeably scattered. It follows that in the interior of atoms there must be small heavy particles. These are called *atomic nuclei*.

We can learn a lot more about the nuclei by accurate observation of the deviated α-particles. For by counting the particles which fly through a bombarded sheet of metal in different directions, we can deduce the laws of interaction of the projectile and the particle struck by it. Direct observation of the path, or at least of its most interesting part, is impossible, for this most interesting part, where the

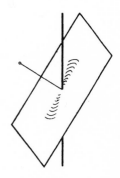

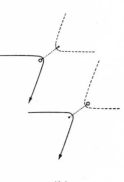

(66)

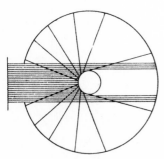

(67)

change from one straight line to another takes place, is certainly of atomic dimensions and hence is invisible even in the Wilson chamber.

We have examples of these collisions in (68) and (69) and Film V. These are drawn according to the laws of ordinary mechanics. This is reasonable enough; for we are dealing with comparatively heavy particles, where, as we know, the quantum effects do not amount to much. We shall come back to this point again (p. 170).

Figure (68) shows the limiting case where the mutual action of the moving particle and the particle at rest is quite trifling until they come within a certain distance of one another, when there is a sudden and large increase in the force. If we think of a sphere described about the stationary particle, whose radius is equal to the distance between the particles at the instant when the repulsive force suddenly rises, the state of affairs is just the same as if very small particles were rebounding elastically from that sphere, like billiard balls (66). In the figure the sphere is shown as a circle. The small particles rebound off it without diminution of velocity, in such a way that their directions after collision make the same angles with the tangent to the circle as they did before collision. We see in (68) that more particles are thrown backwards than forwards. If, however, we carry out the same construction not in the plane but in space, we find that a uniform bundle of flying particles is scattered in all directions.

Figure (69) shows the same phenomenon when

(68)

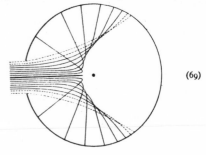

(69)

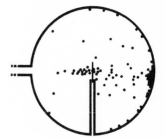

the mutual action of the moving particle and the obstacle, instead of setting in suddenly when they come into direct contact and then disappearing again, extends to some distance. We have in fact assumed that the repulsion is that which acts between two particles with electric charges of the same sign, the so-called Coulomb's law, which has just the same form as Newton's law of gravitation. The force is inversely proportional to the square of the distance; but the force is a repulsion, not an attraction. The paths of the particles are hyperbolas, like the paths of comets, except that with comets the sun lies at the inner focus, whereas with our particles the centre of repulsion lies at the outer focus (67). We see that the directional distribution of the reflected particles is quite different. Every single particle suffers *some* deviation; but for particles which fly past at a considerable distance the deviation is very trifling. Thus by far the greater part of the particles are practically undeviated.

The particles which approach the centre of repulsion closely are scattered vigorously, but by no means uniformly in all directions; many more are scattered forwards than backwards, the fall-off being rapid. This is *Rutherford's law of scattering*.

For another law of force we should have another law of scattering.

If we let a fine beam of α-rays fall on a thin sheet of metal and count the particles deviated in a particular direction, of course it is not one atom that is acting, but many. If the counting apparatus is sufficiently far away, however, this does not matter.

We can then find out by counting the particles whether the scattering takes place in accordance with some assumed law of scattering, or not. The process is shown diagrammatically in Film V. In it a circular screen, to which the particles stick, is shown instead of a counting apparatus. The thickness of the pile of particles immediately shows how the scattered particles are distributed.

Investigations have naturally been made to find out whether the application of wave mechanics makes any difference to the results. As a wave phenomenon the scattering of particles is just like the scattering of light discussed previously (p. 41): a plane de Broglie wave falls on the sheet of metal and sets up a secondary spherical wave at each atom. In the wave theory the field of force surrounding the atom plays a part similar to that played by a condensation of the air in the scattering of light: the wave is to some extent thrown back. The intensity of the secondary waves is different in different directions; according to the statistical interpretation of the waves, this means that the probability of a particle flying off is different in different directions. Calculations have shown that if Coulomb's law holds the wave theory gives exactly the same scattering formula as Rutherford's argument based on the classical mechanics.

This formula of Rutherford's has been amply confirmed by observations: the repulsion is therefore due to electric charges.

Much more, however, can be deduced from these experiments: in fact, they enable us to determine

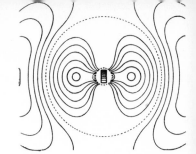

the charge of the nucleus. For we know the charge
and the mass of the projectile: α-particles are helium
nuclei; their charge is equal to 2 unit charges and
their mass is 4 times that of the hydrogen atom.
Now we have seen (p. 92) that an electron which
is shot at a fixed electron at a rate of 10^8 cm./sec.
will approach the latter up to a distance of 10^{-8} cm.
(1 Ångström unit: order of magnitude of the radius
of an atom). It is easy to obtain the necessary modification of these figures when the projectile is an
α-particle and we have a nucleus with several unit
charges, say Z charges, instead of the fixed electron.

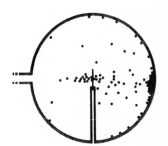

A simple calculation shows that the distance of closest
approach is $r = 1 \cdot 35 \times 10^{-14} Z$. We now refer to
(68) and (69), p. 168, and see that in the hyperbolic
paths the fraction of the α-particles deviated by more
than 90° is about as great as the total number of
collisions in the case of a spherical obstacle, whose
radius is equal to r, the distance of closest approach.
If N is the number of nuclei per c.c., the total area
of the nuclear obstacles is $\pi r^2 N$ sq. cm. Now suppose
there is one deflection for every n incident α-particles
per sq. mm. Then the nth part of a square millimetre
is an effective obstacle, and we have $\pi r^2 N = \dfrac{1}{n}$; hence

$$r = \sqrt{\frac{1}{\pi n N}}.$$

As an example we consider a thin sheet of copper,
bombarded with α-particles in such a way that the
volume affected is a cubic millimetre. The counting shows that the fraction of incident α-particles

deviated by more than $90°$ is about 1 in 10,000, so that $n = 10,000 = 10^4$. There are $N = 2 \times 10^{20}$ copper atoms in a cubic millimetre. We thus have

$$r = \frac{1}{\sqrt{\pi n N}} = \frac{1}{\sqrt{3 \cdot 14 \times 10^4 \times 2 \times 10^{20}}}$$
$$= 4 \times 10^{-13} \text{ cm.}$$

as the distance of closest approach of an α-particle and a copper nucleus. Introducing this in the expression

$$r = 1 \cdot 35 \times 10^{-14} Z, \text{ or } Z = 7 \cdot 4 \times 10^{13} r,$$

we find the nuclear charge of copper,

$$Z = 7 \cdot 4 \times 10^{13} \times 4 \times 10^{-13}, \text{ or about } 30.$$

Other independent determinations of the nuclear charge are obtained by taking angles differing from $90°$ and α-particles from different sources, with different velocities. The experiments all agree in giving the following result, with a considerable degree of accuracy.

The nuclear charge of the element copper, which is number 29 in the periodic table, is approximately 29. (The numerical values, however, are not sufficiently accurate to prove that the nuclear charge is exactly a whole number.)

Similar results are obtained for other elements, leading to this fundamental theorem:

The number of unit charges on the nucleus is equal to the atomic number, that is, the number giving the position of the element in the periodic table (p. 51).

Thus, for example, Chadwick obtained for the

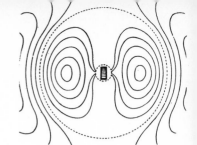

elements copper (29), silver (47), platinum (78), the values 29·3, 46·3, 77·4 respectively for the nuclear charge. It is now clear that the hydrogen atom has a nucleus with a single positive charge, about which one electron is circulating, the helium atom a nucleus with a double positive charge and two electrons, and so on.

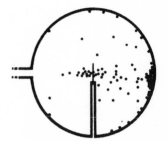

As the electron is only 1/1840 times as heavy as the hydrogen atom, the hydrogen nucleus must contain almost all the mass of the atom. This lightest of all atomic nuclei is called the *proton* (Greek, the first). The same must hold for all the other atoms: the atomic weights are practically the weights of the nuclei. For a long time chemists regarded the atomic *weight* as the characteristic feature of the atom. Now, however, we see that this is not so: it is the nuclear *charge* that determines the position of the atom in the periodic table and hence its chemical behaviour. This result of atomic physics gave an immense impetus to chemistry. A great number of facts and relations which previously were puzzling were suddenly cleared up. In general the atomic weights increase with the atomic numbers. There are, however, some exceptions to this, which are indicated in the table (p. 51) by the symbol ←→. We shall see later that the X-rays confirm the correctness of the arrangement in order of nuclear charge beyond all doubt. A definite answer can also be given to the question whether the periodic table is complete: there are no gaps except those indicated.

Many other and deeper questions, however, now arise.

If the nuclear charge goes up by whole units, must not the nuclei themselves be composite structures, perhaps built up of protons? Then why are their masses, the atomic weights, not whole numbers also?

We shall discuss these and many other questions in Chapter V. For the present we shall not trouble about the structure of the nucleus, but shall take it as an empirical fact that there are as many different nuclei as there are elements in the periodic table. This does *not* reduce one element to another, and it might be thought that all the regularities of the periodic system still remain unexplained. This conclusion, however, is premature. We shall see that the nuclei with their steadily increasing charges collect very remarkable structures of electrons about them, which exactly account for the peculiarities of the periodic system.

In the case of so important a result as the identity of nuclear charge and atomic number, we must look round for independent confirmation of it. This has been obtained in many different ways. For example, it is possible actually to count the electrons, at least in the case of the lighter atoms. In these, the electrons are loosely bound and when subjected to X-rays behave like free particles. Each electron contributes the same amount to the scattering of the X-rays; hence by comparing the intensities of the incident and the scattered radiation we can obtain the total number of electrons in the portion of material used. If we divide this by the number of atoms, we obtain the number of electrons in the

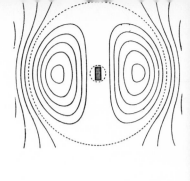

atom. This, as we expected, is found to be equal to the nuclear charge.

Next, how big are the nuclei? In the case of electrons, as we saw on p. 87, we are led to suppose on purely theoretical grounds that they have a finite radius, which is in fact about 10^{-13} cm.; but we have no experimental way of determining this radius. In the case of the nuclei we are in a better position. For small deviations from the Rutherford scattering formula have been detected among those α-particles which are scattered almost directly backwards. We see from (69) (p. 168) that these are the α-particles which strike the atom almost centrally and approach the nucleus most closely. Now we know that the distance of closest approach for direct collision is about 10^{-13} cm. At a distance of this order, deviations from Coulomb's law become noticeable. We can therefore say that the radius of the nucleus is about 10^{-13} cm. Comparing this with the radius of the whole atom, which we estimated from the collisions of gas molecules to be about 10^{-8} cm., we see that the nuclei, like the electrons, are extraordinarily small compared with the atoms which are built up from them; and yet there are grounds for suspecting that the nuclei are themselves composite structures. We shall attempt to penetrate into these deeper recesses of the atom in Chapter V.

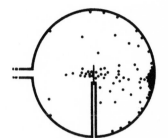

The picture of the atom which we now have is like this: in the interior of every atom there is a positive nucleus, containing practically all the mass of the atom. About it move negative electrons, their number being equal to the number giving the nuclear charge.

To get an idea of the orders of magnitude involved, we may think of a drop of water magnified until it is as large as the earth, the atoms being magnified in the same proportion. The diameter of an atom is then about a metre. The diameter of the nucleus is only about 1/100 mm., that is, the nucleus is a dot scarcely visible to the naked eye. The electrons are about the same size (but several thousand times lighter). These minute dots are circulating about the equally minute nucleus; between them—on the particle theory—there is nothing. The emptiness of intra-atomic space is comparable to the emptiness of the space between the sun and the planets, the distance between the sun and the earth being about 12,000 times the earth's diameter. In the solar system, as in the atom, almost the whole of the mass is concentrated in the central body; but the sun is very much bigger than the planets, whereas the atomic nucleus is about the same size as the electrons.

2. *Bohr's Theory of the Hydrogen Atom.*

The analogy with the solar system was used as a guide in the first attempts to penetrate deeper into the electronic structure of the atom. The simplest atom, hydrogen, consists of the proton with one electron circulating about it—like the earth and its moon. But this resemblance must be merely superficial; for we know that the hydrogen atom, like every other atom, can, only exist in a series of stationary states, a fact which contradicts the fundamental theorems of ordinary mechanics.

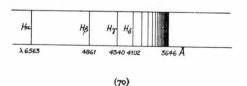

In the case of the hydrogen atom the energy levels or term-values of the stationary states are known very accurately. Balmer has found that a regular series of lines in the hydrogen spectrum (see Plate II(b), facing p. 246, and figure (70)) have the frequencies

$$\nu = R(\tfrac{1}{4} - \tfrac{1}{9}), \ R(\tfrac{1}{4} - \tfrac{1}{16}), \ R(\tfrac{1}{4} - \tfrac{1}{25}), \ \dots$$

where R is a quantity of the nature of a frequency, and is known as the Rydberg constant—after an eminent spectroscopist. Its numerical value is $3 \cdot 29 \times 10^{15}$ vibrations per second.

From the standpoint of Ritz's combination principle we can conclude that the terms of the hydrogen atom are $R/4$, $R/9$, $R/16$, ... Later the term $R/1$ was also found, in a series of lines lying in the ultra-violet. Now, according to Bohr, these terms multiplied by h are the energy levels of the hydrogen atom. As we have explained already (p. 131), these are usually reckoned from the level at which the electron is completely separated from the atom, which is higher than the levels in the bound states. These must therefore be given negative signs. The so-called Balmer terms are then

$$-\frac{R}{1^2}, \ -\frac{R}{2^2}, \ -\frac{R}{3^2}, \ \dots, \ \text{in general} \ -\frac{R}{n^2}.$$

The whole number n was subsequently called the *principal quantum number*; here we shall adopt this name right away.

The arrangement of the terms is shown in (71); the lowest term ($n = 1$) has the value of $-R$, the

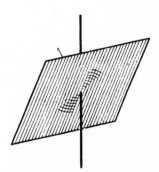

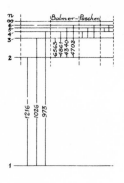

(71)

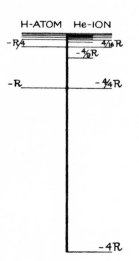

H-ATOM He-ION

(72)

next $-R/4$, and so on. The terms become closer and closer and pile up indefinitely at the energy level zero.

There, where the terms lie extremely thick, the energy is almost continuous. In that region, therefore, ordinary mechanics must hold with fair accuracy. This idea Bohr called the *correspondence principle*, and he applied it with great success. Thus, as early as 1913, he was able to express the energy R and the radius of the orbit of the hydrogen atom, in the ground-state, in terms of the charge (e) and mass (m) of the electron and Planck's constant h.

If the values of these quantities e, m, and h obtained from the observations described above are used, the value of R derived, namely,

$$R = 2\pi^2 m e^4/h^3 = 3 \cdot 29 \times 10^{15} \text{ vibrations per second,}$$

is in excellent agreement with the spectroscopic value given above. The radius of the orbit comes out with the anticipated order of magnitude, namely, that of the atomic radii deduced from the kinetic theory of gases. The exact value is as follows:

Radius of the Lowest State of the Hydrogen Atom
$$= 0 \cdot 532 \times 10^{-8} \text{ cm.}$$

This was the first great success of the quantum theory applied to atomic structure.

A still greater impression was made by the following argument of Bohr. If we take one electron away from a helium atom, it consists of one nucleus and one electron, just like the hydrogen atom, except that the nucleus carries a double charge. Hence the

spectrum of the helium ion must resemble that of hydrogen, but with all the terms four times as large. We should therefore expect the following series of terms:

Hydrogen Atom:
$$-R, -R/4, -R/9, -R/16, -R/25, -R/36, \ldots$$

Helium Ion:
$$-4R, -4R/4, -4R/9, -4R/16, -4R/25, -4R/36, \ldots$$

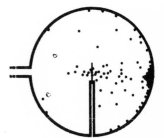

We see that the terms of the first series (indicated by underlining) are all included in the second series.

The same thing is brought out by the figure (72) opposite; every alternate helium term should coincide with a hydrogen term—provided that in the calculation the nuclei are regarded as at rest. Actually the nuclei will move a little; but here, as in the solar system, we have the theorem that the sun and the planet rotate about their common centre of gravity. If we take this into account, the 2nd, 4th, 6th, ... terms of the helium ion will not coincide *exactly* with the 1st, 2nd, 3rd, ... terms of the hydrogen atom, for the masses of the nuclei differ a good deal (the helium nucleus being about four times as heavy as that of hydrogen).

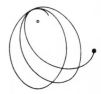

Now a spectrum was actually known, which was obtained from Geissler tubes containing hydrogen and helium, and which contained lines which came exactly between the Balmer lines of hydrogen; hitherto these lines had been ascribed to hydrogen also.

Now Bohr declared, " These are helium lines.

See whether these lines are still obtained when you carefully remove all traces of hydrogen from the tube. If you look closely, you will see that it is a mistake to call these lines Balmer lines of hydrogen, for they are all slightly displaced."

Both assertions were immediately confirmed. This was the second triumph of Bohr's theory.

At that time no one knew anything about electron-waves and there was no such thing as wave mechanics. Physicists had no scruples about applying classical mechanics to the calculation of electronic orbits, even in complicated cases. They obtained orbits whose energy values formed a continuous succession. Additional assumptions had to be made which would enable them to select from this continuous range of energy the series of stationary orbits; these assumptions were called the *quantum conditions.* We shall illustrate their nature only in the simplest case of circular orbits (73). It was found that the correct energy values, the Balmer terms, were obtained by accepting as permissible only those orbits for which the product Momentum × Circumference of Orbit is equal to a multiple of *h*, that is,

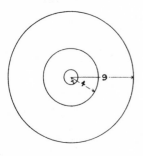

(73)

$$p \times 2\pi r = nh.$$

The product *pr* is called the *angular momentum,* and we say that the angular momentum is " quantized "; that is, it is always a multiple of a certain unit, namely Planck's constant divided by 2π.* The radii of the circles are found to be in the ratios $1 : 4 : 9$ . . . .

*For this constant $h/2\pi$ the notation $\hbar$ has been introduced and will occasionally be used here.

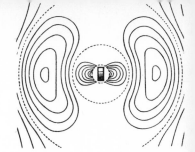

The same condition has been found to apply to all rotations, even in complicated cases. The "whole number" n which appears in the condition above is called the *quantum number*.

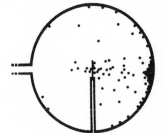

Motions which are not rotations but rather of the nature of vibrations have certain periods. It was found that each of these periods has its own quantum condition. An example is the general motion of an electron in a closed orbit around a nucleus; by Kepler's laws the orbit is an ellipse. This motion, like the circular motion, has only one period, so there is only one quantum condition; we obtain exactly the same simple series of terms $(-R/n^2)$ as we did in the case of the circular orbit. There are an infinite number of ellipses with the same value of n; they all have the same major axis (which is proportional to n), but different eccentricities, their shapes varying from a very elongated form to a circle (74). It is only for the circular orbit that n is proportional to the angular momentum; for the ellipses with the same major axis the angular momentum is less.

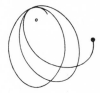

Now it was already known that the Balmer lines do not appear as single lines unless the means of observation are pretty crude; in reality each consists of a group of fine lines. This is referred to as the *fine structure* of the lines.

The energy values, therefore, must possess a fine structure. So one quantum number is not enough to enumerate them with; the motion must have a second period, with a "second quantum number".

Sommerfeld discovered the reason for this com-

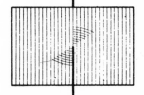

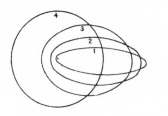

(74)

plication.   In very elongated elliptical orbits the
electron comes very close to the nucleus and reaches
a high velocity there, comparable to the velocity of
light.   Then by the theory of relativity the mass of
the electron is no longer constant, but is slightly
increased. This has the effect of slightly modifying
the whole orbit. The nature of the motion then is
shown in Film VI. The orbit forms a sort of rosette;
the motion can be approximately described as
motion in an ellipse which itself rotates very slowly.
This rotation or so-called *precession* gives the second
period and the second quantum number k.   For if
we consider various ellipses, all having the same n,
that is, the same major axis (one of them being a
circle), the orbital period of the electron is the same
for all the orbits, by Kepler's laws.

The precession of the whole ellipse, however,
gives a new period.   The long narrow ellipses come
closer to the nucleus and hence show the mass-
variation effect more noticeably. The quantum condition associated with the precession is that only
those elliptical orbits are permissible for which the
angular momentum is a whole number multiplied
by $\hbar$, that is, equal to $k\hbar$. This whole number k
is the second quantum number; it is never greater
than n, and is equal to n for the circular orbit.   In
figure (74) a succession of these ellipses for $n = 4$,
$k = 1, 2, 3, 4$ is shown.   Each corresponds to a very
slightly different energy, that is, each term is split
up.

The motion of a particle in space can, however,
have yet a third period, the three periods corre-

sponding to the three dimensions of space. This can also be realized in the hydrogen atom; not when the atom is left to itself, but when it is subjected to an external influence. Such cases were already known to investigators. If a Geissler tube containing hydrogen is excited and placed between the poles of a magnet, each line, as was shown by Zeeman, is split up into several lines—the so-called *Zeeman effect* (Plate III(*e*) and (*f*), facing p. 250). A still more complicated splitting-up occurs in an electric field (the *Stark effect*). Both effects are completely explained by the theory. The field in fact introduces a new period into the motion, corresponding to a new quantum number m. The phenomenon in the magnetic field is easy to describe; the motion is no longer in one plane, but for weak magnetic fields is very nearly so, and may be imagined as like the rosette motion of Film VI, the plane of the orbit rotating slowly about the line giving the direction of the field. From all possible planes the quantum condition selects a small group of planes in the following way:

An angular momentum may be represented geometrically by an arrow, called a " vector ", pointing along the axis of rotation.

Here we have two rotations: that of the electron in its orbital plane (Film VI) and that of the plane about the direction of the field (Film VII). Thus (75) we have to plot the orbital angular momentum as a line of length k on the orbital plane ($\hbar$ being taken as unit), and the magnetic angular momentum as a line of length m along the line of force (through the

(E 969)

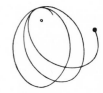

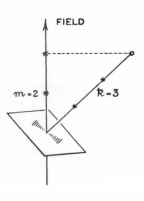

FIELD

$m = 2$ $k = 3$

(75)

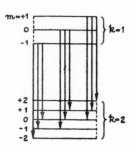

$m = +1$
0
-1
} $k = 1$

+2
+1
0
-1
-2
} $k = 2$

(76)

nucleus). This number m, like k, must be a whole number; it is called the *magnetic quantum number*. As the field is assumed to be weak, the magnetic rotation is slow compared with the rate of rotation of the electron; hence the axis of rotation of the resultant motion coincides almost exactly with the larger angular momentum k. In addition, m must be the projection of k in the direction of the field; hence m can have only the integral values between $+k$ and $-k$.

For example, if

$$k = 0,\ m = \qquad\qquad\qquad 0,$$
$$k = 1,\ m = \qquad\qquad -1,\ 0,\ +1,$$
$$k = 2,\ m = \qquad -2,\ -1,\ 0,\ +1,\ +2,$$
$$k = 3,\ m = -3,\ -2,\ -1,\ 0,\ +1,\ +2,\ +3.$$

This, however, means that the orbital plane can only take up a certain number of positions relative to the direction of the field, the number of positions for $k = 0, 1, 2, 3, \ldots$ being $1, 3, 5, 7, \ldots$.

This phenomenon is called *quantization of direction*. It is one of the most remarkable results of the theory. Later we shall come across a direct experimental confirmation of it (p. 230).

To each position of the orbital plane there corresponds a different energy; hence a splitting of the terms takes place. This, however, is of a very simple type; the distances between the resulting lines are all the same, and the same for every term. Hence the lines of the spectrum should always be split up into three components. This is shown in (76) for the case where $k = 1$ for the upper term, $k = 2$

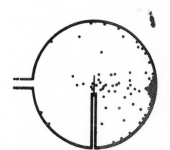

for the lower term. It so happens that in rotations, only those transitions are possible for which *m* does not increase or decrease by more than 1; these are indicated by arrows. There are 9 transitions, it is true, but they form groups of 3, which exactly coincide. We should therefore obtain 3 Zeeman lines. This is actually found on occasion; we then speak of the " normal Zeeman effect " (Plate III(*e*)) —unjustifiably enough, for with most spectral lines the splitting is much more complicated. The reason of this we shall speedily learn.

All these theories depending on classical mechanics correctly reproduce the phenomena actually observed, in their rough outlines, but fail in many details. This is no wonder, for we know that classical mechanics is not really applicable. The quantum conditions, for example, are quite alien to it. On the classical theory, too, the phenomenon of light emission itself remains quite unintelligible. It is described as due to a quantum jump of the electron from one stationary orbit to another; but any further progress is hopelessly barred.

It is clear that all these arguments were of quite a provisional character. They were the steps which prepared the way for the application of the new quantum mechanics to atomic processes. The main features of this new mechanics have already been considered; we know that the electron has wave properties also. We must now see whether these affect its behaviour in the atom, and find out where this idea leads us.

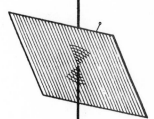

3. *Wave Mechanics of the Hydrogen Atom.*

The real reason why the corpuscular idea is not appropriate for atoms is this: we are interested in the stationary states, which correspond to fixed energy values. If an atom has been observed to be in such a state, then, according to Heisenberg's uncertainty principle, we cannot observe times with any pretence of accuracy, for if the energy is *exactly* known, then the error in the time measurement may be infinite. Hence the course of the motion in time is unobservable, so that we cannot speak of electronic orbits, and our films are a little fraudulent.

The question where the electron is at a given time must therefore be dropped. On the other hand, the *probability* that the electron is at such and such a point is a determinable quantity; for this purpose we have to associate a de Broglie wave with the electron—or better, to replace the electron by the wave and determine the strength of its vibration at such and such a point.

The meaning of the mysterious quantum conditions immediately becomes clear. We again consider a circular orbit and imagine the electron moving in the orbit to be accompanied by a wave. If the radius is of considerable size, we can ignore its curvature and assume that the same de Broglie relation (p. 142) will hold for this revolving wave as for the plane wave, namely,

Momentum $\times$ Wave-length $= h$. $(p \times \lambda = h.)$

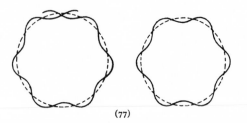

(77)

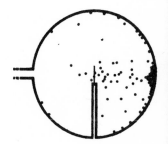

If we now attempt to construct a wave motion along
the circular orbit, with a given wave-length, we do
not in general return, after a complete circuit, with
the same phase of vibration as we started (77). After
a further circuit the phase is different again, and so
on. In short, we cannot assign a definite phase to
each point of the circumference of the circle. We
see at once that this is only possible for particular
circles, namely, when the circumference is an exact
multiple of the wave-length ($2\pi r = n\lambda$). Combining
this relation with de Broglie's quantum rule given
above, we have

$$p \times 2\pi r = p \times n\lambda = hn,$$

or

Momentum × Circumference of Circle
$$= \text{Multiple of } h,$$

which is just Bohr's quantum condition for the case
of a circular orbit. It expresses the condition that
a vibration shall have *one value, and one value only*,
at every point in space.

The state of affairs is just like the elastic vibrations
of a large wheel (without spokes) or of a wire hoop.
In the latter case it is obvious that as each point of
the hoop can only vibrate in one way, the circumference of the hoop must be an exact multiple of
the wave-length. This vibration is very like that
of a stretched piano string, except that the ends are
not fixed but are joined to one another in some way.
In both cases only certain types of vibration are
possible. These are called the *characteristic* or *proper*
vibrations; with the stretched string they are those

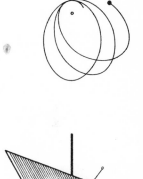

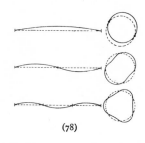

(78)

Number of
Nodal lines.

1

2

3

(79)

for which the length of the string is an exact multiple of half a wave-length, while for the wire bent into a hoop, the length must be a multiple of the whole wave-length (78) (p. 187).

With the string, the fundamental has no nodes, the first overtone one, and so on. With the hoop, the fundamental has two nodes, the first overtone four, the second overtone six, and so on.

Turning back to the electron, we now have to modify our picture considerably; for the vibration is certainly to be imagined as going on in three dimensions. It is comparable not with the vibrations of a hoop, but with those of an elastic substance —something like a table-jelly. But as it is difficult to draw diagrams of these three-dimensional motions, we shall consider a two-dimensional model, namely, the vibrations of a circular membrane, say the parchment of a drum. As is well known, these vibrations can be made visible by strewing fine sand on the vibrating surface. The sand only remains at the places where there is no vibration, along the so-called *nodal lines*, and we obtain the so-called Chladni's figures. The nodal lines are of two kinds, straight radial lines and circles with their centre at the centre of the parchment. The fundamental tone exhibits no nodal lines. We may, however, count the non-vibrating boundary as a nodal line. In the figures the regions bounded by the nodal lines are alternately shaded and white. This is to indicate that the vibrations in the shaded regions are in exactly the opposite phase to those in the white regions; for example, the shaded regions of the

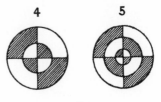

4   5

(80)

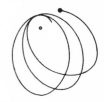

membrane are moving upwards while the white
regions are moving downwards. Here there are just
as many modes of vibration as there are nodal lines
(79), (80).

The capacity of the membrane for vibrating
depends on its being fixed round the boundary. In
the atom, there is no such fixed boundary; instead,
there is the nucleus, which—in particle language
—attracts the electron. This attraction takes the
place of the tension at the boundary.

The mathematical expression of this goes far
beyond the scope of this book. The setting-up of
the so-called *wave equation* is due to Schrödinger,
who took account of the forces acting on the moving
particle. The same equation is the mathematical
expression of collision phenomena, such as Rutherford's experiments on the scattering of α-particles at
nuclei. There, however, we are concerned with
another type of solution of the equation; a plane
wave approaches and is transformed by the action
of the field of the nucleus into a spherical wave.

In the present case, on the contrary, we have to
deal with solutions which die away at a great distance
and give strong vibrations only in the neighbourhood of the nucleus. In this latter region, therefore,
the probability of finding the electron is large; it
decreases rapidly outwards, but theoretically is zero
at an infinite distance only. This infinitely distant
region corresponds to the boundary of the membrane. The frequency multiplied by h is the term-
value or energy of the system in the stationary state.
If the hydrogen atom were all in one plane, we

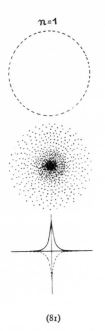

n=1

(81)

n=2

(82)

should have the same vibration diagrams as in the case of the membrane, but its boundary would lie at an infinite distance.

In the actual three-dimensional case we have *nodal surfaces*, nodal *planes* and *spheres*, instead of nodal lines. We shall picture the various modes or types of vibration by means of the nodal surfaces. These pictures, however, are not the only possible ones. For if we have two types of vibration with different nodal surfaces but the same frequency (or energy), we can superpose them and get quite a different picture. We are, however, not really concerned with the shapes of the surfaces at all, but only with their number. We accordingly select a definite configuration of nodal surfaces, which is easy to picture.

We begin by counting the spherical nodal surfaces. As the vibration dies away at a distance, on any very large sphere there will be no trace of it left; hence we can say that the "infinite sphere" is a nodal surface, just as we could regard the edge of the membrane as a nodal line. The number of spherical nodal surfaces, including the infinite sphere, is the *principal quantum number n* in the wave theory. For $n = 2$ we have one finite sphere, for $n = 3$ two finite spheres, and so on.

The annexed figures (81), (82) show the cases $n = 1$ and $n = 2$, diagrammatically in two dimensions, in three ways. The first figure merely indicates by shading how the vibration is divided into portions of opposite phase by the nodal spheres. The appearance of the vibration is also shown by the density

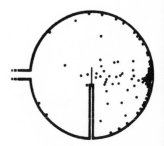

of dots, as seen from above, and in section, showing the extreme amplitudes.   We see that in the
first case the probability of finding the electron
is very great in the neighbourhood of the nucleus
and falls off outwards.   In the second case there is,
in addition, a shell-region of large probability.   For
large values of n further shells occur.   Now it is
found that the radius of the vibrating region in the
case $n = 1$ is about the same as the radius of Bohr's
first circular orbit, the radius of the shell-region in
the case $n = 2$ about the same as Bohr's second
circular orbit, and so on.   It follows that we are
justified in giving n the same name of " principal
quantum number " in both cases, although we are
dealing with things as different as orbits of a particle
and vibrations of a continuous Something.

Each of the spherical modes of vibration may be
accompanied by nodal planes. To classify these, we
take a straight line through the nucleus as axis; then
we have *meridian planes*, which pass through the axis,
and *horizontal planes*, at right angles to the axis (83).

The nodal planes, however, cannot occur in *every*
possible combination, as the nodal lines of a membrane
do. On the contrary, we have the following theorem:

The number of horizontal planes is at most equal
to the number of finite spheres; that is, we have the
following scheme:

for $n = 1$, $l = 0$ only;
for $n = 2$, $l = 0$ or 1;
for $n = 3$, $l = 0$, 1, or 2,

and so on.

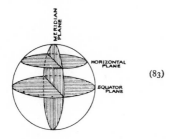

(83)

In Bohr's theory there was a second quantum number k, which was always confined to the values $1, \ldots, n$ and enabled us to distinguish between the various ellipses with the same major axis; when multiplied by h it gave the angular momentum.

Our new l, called, in wave mechanics, the *subsidiary quantum number*, is therefore 1 less than k. The connexion between the angular momentum and the ellipticity of the orbital motion on the one hand and the existence of horizontal nodal lines on the other is not easy to see; we shall subsequently show, in the case of motion in a magnetic field, how angular momenta arise in the wave mechanics. Here, however, we shall content ourselves with the following remark: there is no Bohr orbit without angular momentum; even the orbit of the ground-state is a circular motion, with $k = 1$. In the quantum mechanics, however, there are modes of vibration with no angular momentum; the ground-state $n = 1$, $l = 0$ is one. The subsidiary quantum number gives the angular momentum directly (as a multiple of h).

Figure (84) shows a cross-section of the vibrating system with two spherical nodal surfaces and one plane nodal surface, the amplitude being indicated by the density of dots.

Finally, we come to the meridian nodal planes. Here the various possibilities which may occur are indicated by a quantum number m, which is called the *magnetic quantum number* (p. 184). If there is no meridian plane present, we write $m = 0$. If there is a meridian plane, there are two types of vibration, for which the meridian planes stand at right

(84)

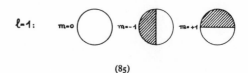

(85)

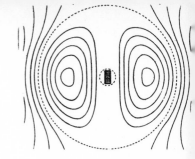

angles to one another; these are denoted by $m = -1$ and $m = +1$.

We have

$$l = 1; \quad m = 0, \, m = -1, \, m = +1;$$

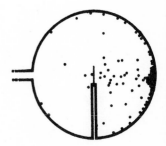

these are the three possible cases if $l = 1$; that is, the plane of the equator is itself a nodal plane, and figure (85) shows how this plane is divided by the meridian planes.

If $l = 2$, we have five cases, which are immediately obvious from (86):

$$l = 2; \quad m = -2, -1, 0, +1, +2.$$

We shall now give a summary showing the number of types of vibration possible with various values of the principal quantum number:

$n = 1; \; l = 0, \, m = 0.$ (1 type.)

$n = 2; \; l = 0, \, m = 0.$
$\qquad \; l = 1, \, m = -1, 0, +1.$ $\Big\}$ $(1 + 3 = 4$ types.$)$

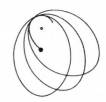

$n = 3; \; l = 0, \, m = 0.$
$\qquad \; l = 1, \, m = -1, 0, +1.$
$\qquad \; l = 2, \, m = -2, -1, 0, +1, +2.$ $\Big\}$ $\begin{matrix}(1 + 3 + 5 \\ = 9 \text{ types.})\end{matrix}$

In general, if the principal quantum number is n there are n^2 types of vibration.

The reason why we have considered the enumeration of types of vibration in such great detail is that on it, as we shall soon see, depends the explanation of the fundamental fact of chemistry, the periodic table.

If we calculate the energies, we find that all the types with the same principal quantum number

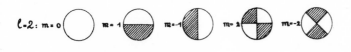

(86)

have the same energy of vibration, and this is equal to the Balmer term $(-R/n^2)$, with the same value R for the lowest term as was previously found by Bohr.

In their very nature the pictures of the revolving electron and of the vibration are so utterly different that it is amazing to find that they give the same energy values. We have seen also that in the case of the collision of a particle with the nucleus, wave mechanics gives the same result as classical mechanics, namely, Rutherford's scattering formula. That these agreements occur is a wonderful piece of luck: otherwise the extraordinary success of modern atomic theory would have been quite impossible.

The agreement goes even farther. If we take the relativistic variation of mass into account, we actually find Sommerfeld's fine structure formula again. The types of vibration with the same principal quantum number n but different horizontal nodal planes then have slightly different energies; this corresponds to the differences between the energies of the quantized ellipses arising from their precession (p. 182). If we imagine the atom brought into a magnetic field, we obtain the same formula as before for the normal Zeeman effect (Plate III(e), facing p. 250). Here the types with different meridian planes give terms with different energy values. This splitting up must correspond to the quantization of direction (p. 184) in Bohr's theory. The latter depended on the rotation and the angular momentum; here, however, we have to do with stationary vibrations—the connexion is not immediately obvious.

We shall go into this matter a little more deeply,

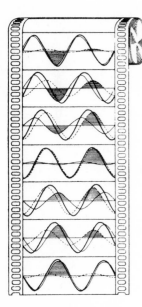

(87)

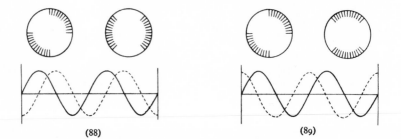

(88)   (89)

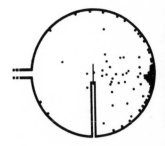

by way of illustrating the correspondence between classical mechanics and wave mechanics. From the diagrams of the meridian nodal planes (85), (86) (pp. 192, 193), we see that there are always two types corresponding to equal and opposite values of m, say $m = -2$ and $m = +2$; these are, at bottom, identical, and differ only in orientation. To make one coincide with the other, we have to rotate one figure until its meridian nodal lines fall exactly where the middle lines of the regions originally were. If we consider, for example, the vibration near the equator only, we obtain figures (88) and (89); one vibration is displaced relative to the other by one-quarter of a wave-length.

This, however, may be brought about in two ways, by rotation clockwise or counter-clockwise. In (88) we have taken clockwise rotation, in (89) counter-clockwise rotation. In the absence of a magnetic field both vibrations have the same energy, i.e. they vibrate at the same rate (as all types of vibration with the same n and l do). Hence their combination also gives a possible vibration. As we see, there are two different resultants, each of which can equally well be taken as representative of the type.

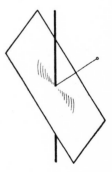

These resultant vibrations, however, are not stationary waves, but progressive (87). The figure shows the two stationary waves with fixed nodes (light and dotted lines) and their resultant (heavy lines), which is a progressive wave. Instead of the stationary waves $m = -2$ and $m = +2$, therefore, we have two opposed revolving waves, which we may denote by $m = 2r$ and $m = 2l$ (r for right, l for left).

Now we have the following state of affairs: although in the absence of a magnetic field the two representations of types of waves are equivalent, in the magnetic field it is not so. Then the stationary waves are no longer solutions of the wave equation at all, but the revolving waves are. The latter have opposite angular momenta and energies, which are displaced equal distances upwards and downwards from the original values.

Here we see clearly how in the wave mechanics the magnetic splitting of terms, the Zeeman effect, again depends on the rotations and angular momenta. In fact, we obtain exactly the same results for the extent of the splitting and the number of terms as we do by the method of quantization of direction which arose from Bohr's ideas of electronic orbits; namely, the normal Zeeman effect (which, however, as we mentioned on p. 185, is in many cases not in agreement with experiment).

The splitting of the terms in the electric field (the Stark effect) can also be calculated; here again the results agree, as regards the positions of the terms, with those given by Bohr's theory.

Wherein, then, lies the advantage of the wave mechanics over Bohr's theory? It seems to come to very much the same thing in the end!

There are some important distinctions, however, which we shall now explain.

The differences of the energy values give the frequencies of the spectral lines. Each line, however, has a definite intensity; there are strong lines and weak lines. On Bohr's theory we can give only

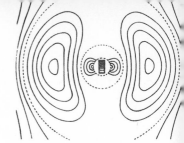

quite rough estimates of the intensity, by calculating
the radiation of the revolving electron on the classical
theory and equating this, according to the correspondence principle, with the energy radiated in
jumps by highly excited quantum states. Quantum
mechanics, on the other hand, gives exact instruc-
tions for the calculation of the intensities of the
spectral lines. These are far too mathematical for
us to give an account of them here; we shall mention
only that this problem was the real starting-point of
the older form of quantum mechanics, the so-called
matrix mechanics of Heisenberg, Jordan, and myself. Later it turned out that Schrödinger's wave
mechanics leads to exactly the same results. These
predictions of the intensity of the spectral lines by
the quantum mechanics, in either theory, go much
deeper than the mere calculation of the term-values;
and they are also abundantly confirmed by experiment.

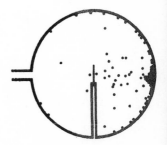

Here is a decisive superiority of the quantum
mechanics over the older methods; but a still more
important point is that the new mechanics forms
a logical, self-contained, self-consistent system—
which could by no means be said of the old theory
—and it has accordingly proved capable of consistent
generalization to atoms with several electrons, and
even to general systems with an arbitrary number
of particles, to molecules, crystals, and metals.

4. *X-ray Spectra.*

The atoms of higher atomic number, which have
more than one electron, present the theoretical

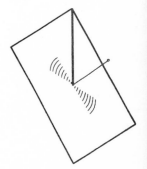

physicist with problems similar to those which the solar system presents to astronomers. In fact, from the point of view of Bohr's theory the resemblance between the two systems is more than superficial. In essentials they present the same problem, except that the comparison is rendered more difficult by the fact that in the atomic case we are concerned chiefly with the forces due to electric charges, whereas in the solar system the forces depend on the masses, and as the mass of the sun outweighs that of all the planets together, the preponderating force is the attraction of the sun, compared with which the interactions of the planets among themselves give rise to small disturbances only. True, the charge of the nucleus is greater than that of any single electron, but its preponderance is not nearly so marked as that of the sun's mass, for the charges of all the electrons taken together amount to the same value (with the opposite sign). Hence the method of perturbations which has been developed in astronomy, on the basis of classical mechanics, yields only rather poor approximations in the case of the atom.

The same is true of the wave mechanics. Here again we can proceed in the first instance as if the electrons had no effects on one another, but were merely connected with the nucleus, and we can subsequently take account of the perturbations arising from the interactions of the electrons.

Mathematically this is even simpler and more elegant than in the classical mechanics. We are not, however, entitled to expect great accuracy, at least without a vast amount of calculation. Hence

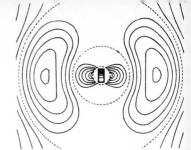

it is important to consider particular cases where
the interactions of the electrons are comparatively
trifling and can be neglected.

In the first place, there are the atoms which have
one easily detachable electron. These we have already
met with; the most important are the atoms of the
alkali metals. They have spectra consisting of series
which closely resemble the Balmer series of hydrogen.
Hence we can assume that one electron is loosely
held and possesses orbits—or vibrations—which are
very like those of the hydrogen atom, while all the
other electrons form a dense cloud round the nucleus
and to a great extent screen its charge from outside
influences. For a nucleus with eleven charges
(sodium), closely surrounded by ten electrons, has
almost the same external effect as a nucleus with one
charge. The spectra of the alkali metals exhibit one
peculiarity, however, which does not fit into our
theory at all; most of the lines are double. We shall
consider this in more detail later (p. 204).

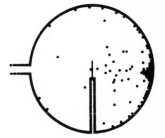

Another case where the nuclear attraction very
greatly outweighs the interaction of the electrons is
the innermost part of all but the lightest atoms.
We have messengers which bring us reports on the
nature of the deep layers of the atom: the X-rays.
In there, close to the nucleus, the laws which hold
are almost as simple as those of the hydrogen atom,
for the motion of the innermost electrons is determined almost entirely by the high charge on the
nucleus.

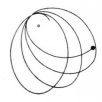

Now for some more exact information about the
X-rays. We have already mentioned (p. 114) that

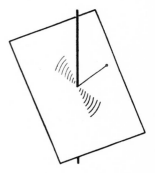

they can be analysed spectroscopically with the help of crystal " gratings ". To investigate the radiation of pure substances we place a plate, called the anti-cathode, which either consists of the substance in question or is covered with a layer of the substance, in the X-ray tube, facing the cathode. It is found that, in addition to a continuous spectrum, single lines are obtained, just as with gases. Each kind of atom has its own characteristic lines. Those of shortest wave-length are called K lines, those of somewhat longer wave-length L lines, M lines, and so on.

The difference between these spectra and optical spectra is as follows: the optical spectra of two " neighbouring " atoms, that is, atoms whose nuclear charges differ by 1, differ very widely, and the similarity between the spectra of atoms of the groups of elements which stand in the same vertical column of the periodic table (such as the alkali metals) is only very rough. The X-ray spectra, on the other hand, are qualitatively the same throughout long series of neighbouring elements, the lines merely being relatively displaced (Plate V(*b*), facing p. 258). This shows that they arise from the interior region of the atom, where the electronic structure of all atoms is very similar, differing only in its extent and the strength of the bindings. The fact that the vibrations of X-rays are many thousand times faster than those of the lines of optical spectra is further evidence for this assumption.

The combination principle holds for X-ray spectra just as it does for optical spectra; there are certain

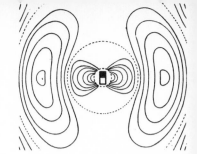

"terms", whose differences are the frequencies. There is, however, one point of difference: there are no absorption lines, the absorption setting in suddenly at a definite point in the spectrum, the so-called *absorption edge*, and extending from there in the direction of increasing frequency (Plate IV(*b*), facing p. 254). Within this continuous absorption band there are further absorption edges, where the absorption suddenly rises.

The essential point is that the frequencies of the absorption edges of an atom are found to give at the same time the terms of the emission lines.

This empirical fact has given rise to an interpretation of X-ray spectra which is also evidence for the most important feature of atomic structure: the arrangement of the electrons in shells. How this is explained theoretically we shall see later (p. 218). Here we are concerned with a convincing interpretation of the empirical relations, due to Kossel:

The electrons arrange themselves in shells. Some of them are held very fast; these form the K shell. Then follow electrons which are more loosely held, forming the L shell, the M shell, and so on, in order.

If light falls on the atom, it may knock out an electron. Ordinary light with a small quantum can only overcome the binding of the outermost electrons: this is the ordinary photo-electric effect (pp. 93, 118). X-rays, however, are capable of knocking out inner electrons. There is a certain minimum quantity of energy required to break the binding of a K electron, an L electron, an M electron, and so on; this at once explains the existence of the absorption edges. We

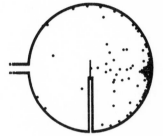

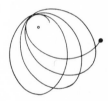

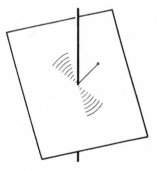

can now label these: that of shortest wave-length (greatest frequency) is the K edge, the next the L edge, and so on.

If, however, an electron is removed from the K shell by the impact of light or of an electron, there

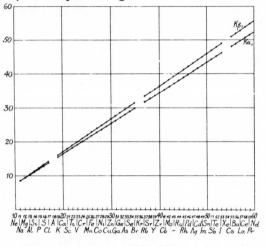

Diagram of points with abscissæ equal to the atomic number Z, and ordinates equal to the square root of the frequencies of K lines

(90)

is a gap left in that shell, and an electron from another shell can now fall into that gap. This leaves a gap in its original shell, say the L shell. The energy difference between the K absorption edge and the L edge is emitted as an X-ray photon. Thus the main properties of the X-ray spectra are explained.

The term-values are determined by analysis of the emission spectra much more accurately than

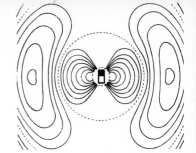

they can be from observations of the absorption edges.

According to Moseley, the data thus obtained serve to check the correctness of the order of the periodic table, and to settle whether the table is complete or not. We have already indicated the possibility of doing this. If, for example, we plot the K terms for successive elements, they advance regularly, as is shown opposite in (90). To an increase of the nuclear charge by 1, there always corresponds the same increase in the term. If an element were missing, the smoothness of the curve would be broken. This would also happen if two elements were in the wrong order. Thus we can verify that at the points in the periodic table which we marked by ←→ (p. 51) it is actually the nuclear charge, and not the atomic weight, that decides the proper position of the elements.

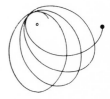

Moseley noticed that the curve of the K terms is just the ground term of an atom having one electron but a nucleus with more than one charge, calculated according to the formula for the hydrogen atom. Sommerfeld then showed that for heavier elements the higher terms of this model give the corresponding curves for the L term, the M term, and so on. Physically this means that the innermost shells of electrons are affected chiefly by the nucleus. It is astonishing how fine details of X-ray spectra can be explained in this way. The most important fact, however, is that Sommerfeld's formulæ hold for all atoms, right up to the last, uranium, which is number 92. The quantum mechanics is therefore

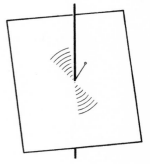

capable of correctly representing an extensive range of systems, differing widely in magnitude and strength of binding.

5. *The Spin of the Electron.*

It looks as if we now had the foundations of a comprehensive theory, and by systematically applying it should be able to explain the structure and the properties of the electronic shells of all atoms.

This assumption, however, is premature. We have already indicated more than once that the results of wave mechanics are only approximately, not exactly, in agreement with the facts.

An important case is the doublet nature of the lines in the alkali spectra. If a little common salt is brought into a hot flame, e.g. that of the gas stove, it colours the flame yellow. Common salt is a compound of chlorine and sodium; in the heat of the flame it is decomposed and sodium atoms become excited and emit the yellow light. The sodium spectrum contains a brilliant yellow line. It is one of a regular series, the remaining lines of which are less striking. Even with a small spectroscope we can see that the yellow line is double. The same line, by the way, is present in the sun's spectrum as a dark absorption line, and is there denoted by the letter D; its occurrence demonstrates the presence of sodium vapour in the outer layers of the sun's atmosphere.

Most of the lines, not only of sodium, but also of potassium and the other alkali metals, exhibit the same doublet structure.

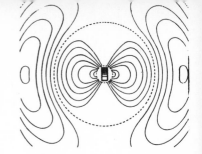

Now there is not the least possible doubt that all these lines arise from the jumps of an outer electron. We have already mentioned the reasons for assuming this: the ease with which the atom is ionized, the ordering of the lines in series with regularities similar to those of the hydrogen atom, and so on.

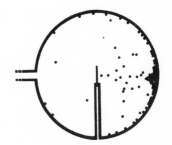

The doubling of the lines involves a new quantum number. In addition to their current number, the terms of a series require a further distinction, that between the first and second terms of the doublet. In the case of the D lines of sodium we write D_1 and D_2.

In the language of Bohr's theory this quantum number, with its two values 1 and 2, implies the existence of a new period; in the language of wave mechanics, it implies the existence of a new class of nodal surfaces.

We have seen, however, that in the case of the hydrogen atom all the possibilities with regard to periodicity (or nodal surfaces, in the wave picture) are used up (pp. 182, 192), and the same is true for all cases of motion of a single electron.

At first it was thought that the new period arises from interaction with the other electrons, which are close to the nucleus. This idea, however, would not work. Besides, in other cases, in more complicated atomic spectra and in X-ray spectra, indications of similar doublets, triplets, and so on were found, none of which could be brought within the theory as hitherto developed. Something essential must therefore be wanting.

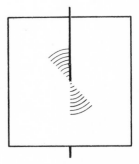

This missing factor must yield a new period, and that even when there is only a single electron present. So long as the motion of the electron is described by stating the position of its centre, the motion can have only three periods (in wave mechanics, three kinds of nodal surfaces). Hence this description is inadequate.

To explain this multiplet structure of spectral lines, Uhlenbeck and Goudsmit put forward the idea that the electron is an entity which not only moves forward as a whole, but rotates on an axis like a top. This internal rotation is the new periodic motion; it is called the *spin of the electron.*

(91)

If anything is going to rotate, it must, according to ordinary ideas, possess extension in space. In the case of the electron, physicists have as far as possible avoided speculations about its size and dimensions. Such speculations lead to difficulties, of which we have already spoken (pp. 73, 87). Accordingly, it has become customary to think of an electron as a point-charge with an arrow attached to it (91), the arrow indicating the angular momentum, which, of course, is plotted as a vector along the axis of rotation, as we saw on p. 183. To help the imagination one can add a circle with an arrowhead, in the equatorial plane of rotation. But one should not imagine that there is anything of the nature of matter actually rotating. The idea of a spin without the existence of something spinning seems to be rather abstruse. But one should remember that there are other examples of such abstractions; for instance, the theory of relativity has deprived the ether, the carrier of electro-

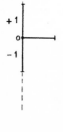

(92)

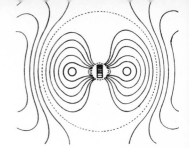

magnetic waves, of all properties of ordinary matter, so that one has to speak about vibrations without having anything material which vibrates.

Every angular momentum exhibits the phenomenon of quantization of direction. If we employ the construction explained in connexion with Bohr's theory (p. 184), the least number of positions which an angular momentum can take up is 3, that is, when it has its own least value of 1 (in h units). For an interval of length 1 can be projected in only three ways on a given line so that the projection has an integral value (92); either end on, so that the projection is 0; or upwards, so that it is $+1$; or downwards, so that it is -1. If, then, the electron were to behave like a top with angular momentum 1, under the influence of the nucleus and the remaining electrons it could set itself in three positions. Each term would therefore be split into three.

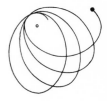

In the one-electron spectra of the alkali metals, however, we actually have a splitting into only two terms. The spin is therefore not an ordinary angular momentum quantized in whole numbers.

The correct duplication is obtained as follows. The projections of the spin in the direction of the outer field must always, as hitherto, have 1 for their distance apart, but they are to be only two in number. This is only possible if we ascribe the magnitude $\frac{1}{2}$ (times $\hbar$, of course) to the spin. For the spin $\frac{1}{2}$ can set itself only in the direction of the field or in the opposite direction; the zero position is excluded, because it would be only at the distance $\frac{1}{2}$ (not 1) from the ends (93). Accordingly,

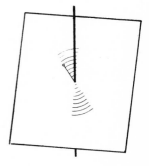

$$\frac{1}{2} \left.\begin{matrix} \\ \end{matrix}\right\} 1 \qquad (93)$$
$$\frac{1}{2}$$

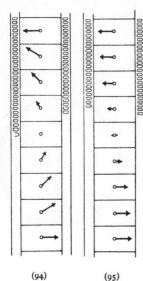

we have a spin quantum number s with the values $+\frac{1}{2}$ and $-\frac{1}{2}$.

The translation of this assumption into the language of wave mechanics depends on the following idea, suggested by the analogy with light-waves.

Light-waves are electromagnetic vibrations. At each point of space over which such a wave passes, the electric and magnetic forces are oscillating to and fro. If we pay attention to the first only, the force at any instant can be represented by an arrow whose direction and length are continually altering.

In figure (94) snapshots of this arrow are drawn beside each other; if they were drawn one after another on successive pages, they would give a " film " of the motion of the arrow. It may happen that the arrow vibrates in one direction only, as shown in figure (95); we then say that the light wave is " polarized ". Experimentally, polarized light is produced by making ordinary light pass through certain crystals; but we cannot go into this here. There are two kinds of polarized light, vibrating in directions at right angles to one another; the two vibrations are indicated (97) by the double arrow (the observer being supposed to be looking in the direction opposite to that in which the light is travelling).

Now the spin may be regarded as a sort of *polarization of the de Broglie wave*. Hitherto we have treated the latter as analogous to a sound wave, in which condensations and rarefactions of the air vary periodically; that is, as an undirected phenomenon. The spin, how-

(94)   (95)

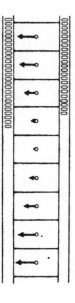

(96)

(97)   (98)

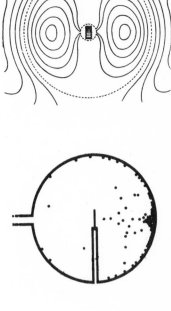

ever, may be taken account of by assuming that, in the de Broglie wave also, that which is vibrating has a direction. The distinction is only this, that in the electric wave of light the force-arrow is vibrating up and down, its direction continually alternating, while in the de Broglie wave the spin-arrow retains its direction, and its magnitude alone varies periodically, to one side or the other according to the direction of spin (96). Looked at from the front, therefore, there are two single arrows in opposite directions (98). If a de Broglie wave is totally polarized " upwards ", this means, in corpuscular language, that the spin angular momenta are all pointing upwards; similarly for " downwards ".

Just as we can have unpolarized or partially polarized light, however, we can also have un-polarized or partially polarized de Broglie waves. These mean a beam of electrons in which some of the electrons have their spins in one direction and some in the other. Each such wave is a combination of one or other of the two types of wave in varying proportions. This proportion means the relative probability of finding, in the beam, electrons with their spin upwards or downwards.

These relations are expressed mathematically by replacing the single Schrödinger wave equation by a system of several equations. These were first stated by Pauli and were then generalized by Dirac in accordance with the theory of relativity. The difference is similar to that between the single wave equation for sound and Maxwell's equations for the electromagnetic field.

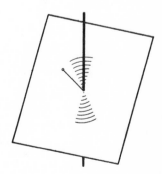

Many features of spectra, doublets, triplets, and so on, as well as the anomalies of the Zeeman effect, are made intelligible with the help of this theory. Yet it is not sufficient; an essential idea is still wanting, to which we shall now direct our attention.

6. *Pauli's Exclusion Principle.*

When a theory works qualitatively, but not quantitatively, the position is unpleasant, but not hopeless. Perhaps we have forgotten to take some small effect into account. If, however, a theory will not even work qualitatively, some error of principle must be involved. Such a failure occurs, for example, if the theory predicts a number of spectral lines differing from the number found experimentally. Numbers, I mean *whole* numbers, are strict judges; they cannot be bribed into agreement by slight alterations. Three remains three and four remains four, their difference is one, no less, no more.

The evil symptoms of such a failure of the theory appear even when we try to construct the next simplest atom to hydrogen, namely, helium. We have a doubly charged nucleus and two electrons, and think of the latter as added one at a time. If we begin by taking one electron only outside the nucleus, we get the same stationary states as we do for the hydrogen atom; we obtain the helium ion, which we have already discussed (p. 179). We express the lowest states by the values of the quantum numbers (n_1, l_1, m_1) of the first electron, in addition to which we also have the spin quantum number s_1:

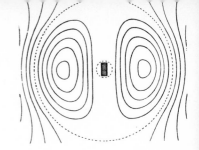

(a) $n_1 = 1$, $l_1 = 0$, $m_1 = 0$, $s_1 = -\frac{1}{2}$,
(b) $n_1 = 1$, $l_1 = 0$, $m_1 = 0$, $s_1 = +\frac{1}{2}$,
(c) $n_1 = 2$, $l_1 = 0$, $m_1 = 0$, $s_1 = -\frac{1}{2}$,

. . . .

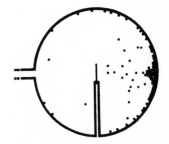

If the helium ion is now transformed into a neutral atom by the addition of another electron, the orbits or vibrations of the two electrons will, it is true, be considerably distorted by their interaction; but qualitatively we must obtain the right result, that is, the right *number* of terms, by first ignoring the interaction and thinking of the second electron as subject only to the influence of the nucleus. Subsequent consideration of the mutual perturbation can only displace the terms, not alter their number.

Hence to the second electron we must assign the states:

(α) $n_2 = 1$, $l_2 = 0$, $m_2 = 0$, $s_2 = -\frac{1}{2}$,
(β) $n_2 = 1$, $l_2 = 0$, $m_2 = 0$, $s_2 = +\frac{1}{2}$,
(γ) $n_2 = 2$, $l_2 = 0$, $m_2 = 0$, $s_2 = -\frac{1}{2}$.

. . . .

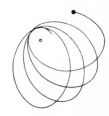

The whole system then possesses states which are obtained by combination of each state of the first set with each of the second; these may be denoted by

$$(a, \alpha), \ (a, \beta), \ (b, \alpha), \ (b, \beta),$$

and so on. We shall consider these four only. Now it is clear that in the absence of a magnetic field the states (a, α) and (b, β) will be identical, and likewise (a, β) and (b, α); for obviously it is only a question of whether the two spins are parallel (both $+\frac{1}{2}$ or

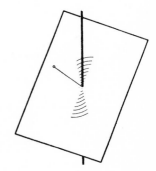

both $-\frac{1}{2}$) or antiparallel (one $+\frac{1}{2}$, the other $-\frac{1}{2}$).

We should therefore expect to get two lowest states, with principal quantum numbers $n_1 = 1$, $n_2 = 1$. All other states, where at least one of the principal quantum numbers is equal to 2 or more, will lie much higher; in the case of hydrogen, in fact, the Balmer terms for $n = 1$ and $n = 2$ are in the ratio of $4 : 1$, the lower corresponding to a frequency in the remote ultra-violet, the higher to one in the visible region. There is, therefore, no possibility of confusing these two lowest terms with any others.

Experiment, however, gives the clear and definite result that only one of the two terms appears, namely,

$$(a, \beta) = (b, a),$$

while the other, $(a, a) = (b, \beta)$, is missing! The number of terms is not correct, which means that we have overlooked some fundamental law.

Pauli thereupon investigated spectra systematically and discovered a great number of similar cases. He also recognized the principle which lies at the root of the matter!

The terms where all the four quantum numbers of two electrons are exactly the same are always missing, and two terms which are formally obtained from one another merely by interchanging the values of all the quantum numbers of two electrons in reality form only one term.

In actual fact, for the $(a, a) = (b, \beta)$ term of helium the two electrons both have the quantum

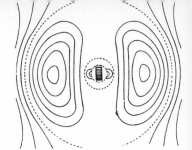

numbers $n = 1$, $l = 0$, $m = 0$, and either $s = +\frac{1}{2}$ or $s = -\frac{1}{2}$; this term is actually missing. In the term $(a, \beta) = (b, a)$, however, n, l, m are equal, it is true, but s_1 and s_2 are different.

As an example of the second part of the theorem we may adduce the terms

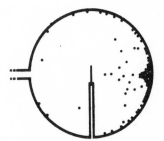

$$n_1 = 2, \ l_1 = 0, \ m_1 = 0, \ s_1 = \tfrac{1}{2},$$
$$n_2 = 2, \ l_1 = 1, \ m_1 = 0, \ s_1 = \tfrac{1}{2},$$

and

$$n_1 = 2, \ l_1 = 1, \ m_1 = 0, \ s_1 = \tfrac{1}{2},$$
$$n_2 = 2, \ l_2 = 0, \ m_1 = 0, \ s_1 = \tfrac{1}{2},$$

which are in reality only one term.

What does this mean?

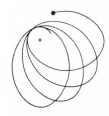

It means that in associating quantum numbers with the individual electrons we were pushing the corpuscular idea too far; in doing this we assumed that the electrons are *distinguishable*, like individuals whom one can call by name. Such is obviously not the case. We get a deeper understanding of the situation by using the wave idea. Waves obviously have no individuality; their job is merely to say, " Here there is such and such a probability of finding a particle: which of the two identical particles it is, we do not know ".

This principle of Pauli's has far-reaching consequences. Firstly, it brings the theory of spectra, of the Zeeman effect, and of many similar phenomena into perfect agreement with observation.

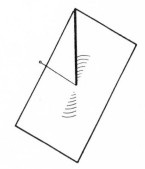

Secondly, and this is a cruel blow, it leaves what we started this book with in more or less complete confusion—namely, the foundations of statistics, which we used in the kinetic theory of gases.

To keep on safe ground, we shall now speak not of an ordinary gas but of an " electron gas ", such as is supposed to exist in the interior of metals and to account for their high conductivity. According to our former principles, we should have to give each of the electrons a name, Edward, John, George, and so on, and find out in how many ways a definite grouping, say a distribution between the halves of the available space, can arise. In Film I we accordingly drew each atom differently, so that the individual paths of the separate electrons could be followed more easily by the eye. From this counting of frequency we then obtained the most frequent distribution as the state of equilibrium, and hence the whole kinetic theory of gases.

Is all this wrong, then? Undoubtedly: if the electrons are *not* individuals, if they are utterly indistinguishable, the counting of their " arrangements " by the old method is wrong.

Fermi and Dirac, however, were not deterred by this, but carried out the enumeration according to the new method. For the spatial distribution this gives nothing of particular interest; but if we ask with what frequency we are to expect states of the electron gas in which there is a definite distribution over the possible energies of the whole gas, something quite new comes out.

The extremely satisfactory discovery was made that the application of the new Fermi-Dirac statistics to the electrons in metals gives much better results than the older theory, a great many previous diffi-

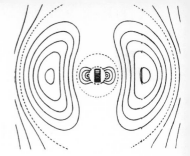

culties being removed.   Indeed, it was only then that a practicable theory of electrical and magnetic phenomena in metals became possible.   This is a new line of evidence for the Pauli principle.

This new statistics, however, cannot be applied to atoms or molecules, for these, of course, contain a *number* of electrons, each of which has a spin. Hence they behave quite differently from single electrons.   The necessary modification was made by Einstein, with the help of a method originally devised by Bose for a gas consisting of photons (light quanta); but owing to the great mass of the atom compared with that of the electron, deviations from the ordinary behaviour can only be expected to occur at temperatures so low as to be practically inaccessible.

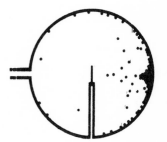

The most important application of the Pauli principle is in the interpretation of the periodic properties of the elements.

7. *The Meaning of the Periodic System of the Elements.*

We want to understand why substances are what they are.   We know that their great variety depends on the fitting together of 92 kinds of atoms to form all manner of structures.   These 92 different kinds of atoms consist of 92 different nuclei with corresponding swarms of electrons.   The nature of the chief problem with which we are faced is made clear if we contrast the following results, which we are driven to accept.

The nuclei form a simple series, the charge increasing step by step by the same amount, the

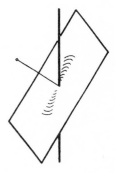

elementary quantity of electricity (p. 88). (The masses increase likewise, but not quite so regularly.)

The shells of electrons, however, behave in quite a different way. True, the number of electrons also rises regularly by unity from element to element. Nevertheless, none of the properties of the atoms, physical or chemical, change in this regular way, but they show a periodic behaviour, first with the period 8, then with the period 18, and finally with the period 32—although there are traces of the period 8 showing all the time.

How is it possible for the manifold variety of atoms and their periodicities to arise from the regular increase of the nuclear charge and the addition of the corresponding electrons?

The kaleidoscopic aspect of the material universe is due, in the last resort, to the periodic behaviour of the elements. So the solution of our problem is really a matter of pressing importance.

The nature of the problem may be readily grasped from the following analogy:

There are savage tribes in Africa who use cowrie shells as their currency. If one of them buys an egg, he pays, say, 10 cowries for it; if he buys a hen, perhaps he has to pay a coconutful of cowries, and if he wants to take a wife, his father-in-law will demand a few sackfuls of cowries.

Civilized peoples have replaced this complicated arrangement by a coinage system. If we buy an egg, we pay a penny; for a hen we pay half a crown, and so on.

Here is an ideally simple coinage system:

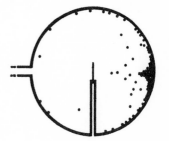

Copper coins for 1, 2, 5 pence;
    10 pence = 1 shilling.
Nickel coins for 1, 2, 5 shillings;
    10 shillings = 1 crown.
Silver coins for 1, 2, 5 crowns;
    10 crowns = 1 sovereign.
Gold coins for 1, 2, 5 sovereigns;
    10 sovereigns = 1 (?).

Our actual coinage is not by any means so simple as this; but the principle is essentially the same. The coins form a periodic system. Although their purchasing power steadily rises, the number stamped on them always recurs after four steps.

What is the reason for this periodicity? Obviously, the desire not to have our pockets and purses weighed down by excessively large pieces of metal. This is an "exclusion principle"—and here comes the analogy with Pauli's principle for the electrons in the atom (p. 210). Analogies, however, must not be pushed too far; we shall now drop this one and consider how the Pauli exclusion principle gives rise to the periodicity of the systems of electrons.

We begin with the hydrogen atom and replace the singly charged nucleus by a doubly charged one, that of the helium atom, as we did in last section (p. 211). We saw there that the newly added electron must take its place in a perfectly definite lowest state; for as the quantum numbers n, l, m are the same for both electrons ($n = 1$, $l = 0$, $m = 0$), the spins must be antiparallel ($s_1 = \frac{1}{2}$. $s_2 = -\frac{1}{2}$).

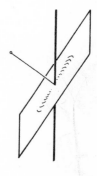

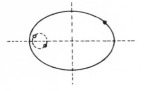

We now replace the helium nucleus by the triply charged lithium nucleus and add a third electron. Then this certainly cannot be in a state with the same quantum numbers $n = 1$, $l = 0$, $m = 0$, $s = +\frac{1}{2}$ or $-\frac{1}{2}$; there are only two such states. These two states form the innermost shell of this atom and all subsequent atoms; X-ray spectroscopists call it the K shell (cf. p. 201).

As l must always be less than n, and m is at most equal to l, the next electron must go into a state for which $n = 2$. Figure (99) shows the configuration of the three electrons in the form of Bohr's orbits. The two K electrons screen the nucleus and reduce the external effectiveness of its charge nearly to 1. Hence the third electron is much more loosely held than the K electrons. The third element, lithium, does actually exhibit the properties which we are thus led to expect; it belongs to the group of alkali metals and has an electron which is relatively easy to detach. This forms the start of the next shell, which in the X-ray connexion is called the L shell.

If we continue this method of step-by-step increase of the nuclear charge with addition of electrons, the L shell will fill up in the first place. How many electrons can it accommodate in all?

Just as many as there are states for which $n = 2$. Now we have seen (p. 193) that the hydrogen atom has 4 such states, electron-spin being neglected; namely, the state $l = 0$, $m = 0$, and the three states, $l = 1$, together with $m = -1$, 0, or $+1$. There is still the spin, which can have one of two values;

(99)

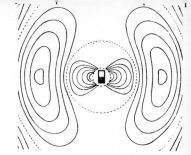

each electron can have the spin $+\frac{1}{2}$ or $-\frac{1}{2}$. Thus we obtain at most $2 \times 4 = 8$ electrons in the L shell.

This is the actual extent of the first period:

$$\text{Li} \quad \text{Be} \quad \text{B} \quad \text{C} \quad \text{N} \quad \text{O} \quad \text{F} \quad \text{Ne.}$$

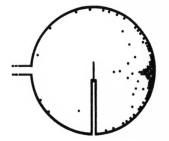

When neon (Ne) is reached the L shell is " full up "; there is no room for any more electrons. The structure forms a completed whole, a fact which at once explains why neon is an inert gas, having no inclination to interact with other atoms, that is, being chemically inert. The preceding element fluorine, however, just needs one electron to complete the L shell; this explains why fluorine readily appears as a negative ion, the gap being readily filled with an electron.

In the same way, beryllium (Be) tends to occur as a divalent positive ion (losing its two outer electrons) and oxygen as a divalent negative ion (filling two gaps and completing the L shell). Carbon, in the middle, has a dual character; it can be stripped down to a helium-like structure or built up to a neon-like structure. This intermediate position is one of the reasons why it enters into so many compounds, the whole of organic chemistry being just the chemistry of the compounds of carbon.

Further addition of electrons is only possible if $n = 3$; we then get the M shell. How many electrons can it hold?

Firstly, we can have $l = 0$ or 1; along with $s = +\frac{1}{2}$ or $-\frac{1}{2}$, this again gives the same 8 possibilities as in the L shell. In actual fact, the

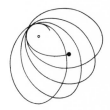

next row of the periodic system begins with 8 elements,

$$\text{Na} \quad \text{Mg} \quad \text{Al} \quad \text{Si} \quad \text{P} \quad \text{S} \quad \text{Cl} \quad \text{A},$$

which in every respect, physical and chemical, resemble those of the preceding row. But this does not complete the M shell. For if $n = 3$, we can have $l = 2$; this gives the 5 possibilities

$$m = -2, \ -1, \ 0, \ +1, \ +2,$$

and as in each case the spin can be either $+\frac{1}{2}$ or $-\frac{1}{2}$, there are 10 further places in the M shell. The total number of places is therefore $8 + 10 = 18$, or, better, there are at most $2 \times 9 = 18$ electrons in the M shell; for $3 \times 3 = 9$ is the number of states of the hydrogen atom for $n = 3$, and the 2 comes from the spin.

It is clear that there can be at most $2 \times 16 = 32$ electrons in the N shell—in agreement with facts.

We need not follow out the details any further, except as regards a few points. Thus, the order of addition of new electrons is not always the same as at the beginning of the periodic table, where one shell is completely filled before the next is begun. The process goes on regularly up to element number 19, potassium (K); from this point the two outermost shells, M and N, compete for electrons. The N shell never has more than one or two electrons until the (inner) M shell is filled up. All these elements, Ca, Sc, Ti, V . . . after A therefore resemble one another in having two loose electrons (or one): see Table II. A similar situation sub-

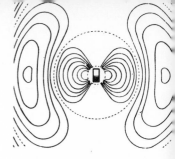

TABLE II

DISTRIBUTION OF THE ELECTRONS IN THE LIGHTER ATOMS

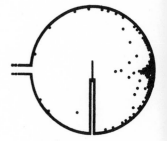

Element	K $n = 1$	L $n = 2$	M $n = 3$	N $n = 4$
H 1	1	—	—	—
He 2	2	—	—	—
Li 3	2	1	—	—
Be 4	2	2	—	—
B 5	2	3	—	—
C 6	2	4	—	—
N 7	2	5	—	—
O 8	2	6	—	—
F 9	2	7	—	—
Ne 10	2	8	—	—
Na 11	2	8	1	—
Mg 12	2	8	2	—
Al 13	2	8	3	—
Si 14	2	8	4	—
P 15	2	8	5	—
S 16	2	8	6	—
Cl 17	2	8	7	—
A 18	2	8	8	—
K 19	2	8	8	1
Ca 20	2	8	8	2
Sc 21	2	8	9	2
Ti 22	2	8	10	2
V 23	2	8	11	2
Cr 24	2	8	13	1
Mn 25	2	8	13	2
Fe 26	2	8	14	2
Co 27	2	8	15	2
Ni 28	2	8	16	2
Cu 29	2	8	18	1
Zn 30	2	8	18	2
Ga 31	2	8	18	3
....	..	..	..	..

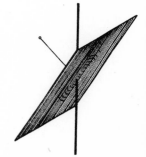

sequently recurs several times. Hence, for example, arises the superficial impression that the group of the so-called rare earths does not really fit into the periodic table. Actually, however, the theory explains why these peculiar elements should occur at all. They arise because the two outermost shells, the O shell ($n = 5$) and the P shell ($n = 6$), remain unaltered, always containing the same number of electrons, while an inner shell, the N shell ($n = 4$) is filled up subsequently. As a result, these elements resemble one another so closely that their separation was a great chemical achievement.

This interpretation of the periodic system, an interpretation due in essence to Bohr, has given a powerful impetus to chemistry. Indeed, we may say that the distinction between physics and chemistry has disappeared, so far as theory is concerned; the difference is merely one of practical methods and modes of instruction. For even the nature of chemical forces has had light thrown on it by the quantum theory.

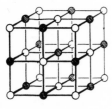

(100)

We have already seen that the tendency of the alkali metals Li, Na, K, Rb, Cs to occur as positive ions is just as intelligible as the tendency of the halogens F, Cl, Br, I to form negative ions. As opposite charges attract one another, it is no wonder that a sodium ion and a chlorine ion join forces to form a molecule of NaCl, the substance we call common salt. This kind of electrical binding explains most of the properties of compounds of the salt type quantitatively.

These salts consist of crystal lattices, in which the

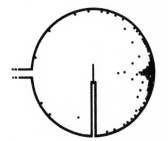

two kinds of ions are arranged alternately, like the squares of a chess-board (100). Each ion is surrounded by six ions of the other kind, but we cannot say that any two ions are associated with one another in any special way. There are no real molecules, therefore; or rather, we may say that the whole crystal is a single gigantic molecule. This kind of combination seems quite natural when we explain it by electrical forces. For the forces coming from an ion extend in all directions, and though they catch an ion of the other kind, this does not weaken them and they are capable of attracting further ions.

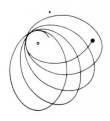

There are other types of compound, however, which behave quite differently. Chemists speak of the " saturation " of the binding forces or " valencies ". The simplest case is that of two identical atoms, as in the gases hydrogen, oxygen, nitrogen. As a rule the chemist expresses these by the symbols H_2, O_2, N_2, or, if he wants to be very accurate, by H—H, O=O, N≡N, the dashes being intended to indicate the forces holding the atoms together. Each atom is thought of as having one or more little hooks, indicated by dashes: H—, O=, N≡. Chemical combination arises from the fastening together of the hooks: H— and —H give H—H. Any other hydrogen atom meeting this molecule finds no free hook and is ignored; there is no such thing as H_3. O= and =O give O=O, with a double bond. If, however, another O= comes along,

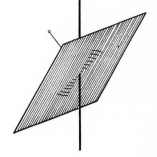

it can burst the double bond and form $\overset{\displaystyle O}{\underset{\displaystyle O—O}{\diagup\diagdown}}$; this is ozone, O_3.

It is very important that we should be able to explain this phenomenon of valency saturation. The simplest case, that of two hydrogen atoms, exhibits all the essential features. There we have two nuclei, each with one electron. When the atoms approach one another very closely, the electrons no longer remain beside their own nuclei all the time, but begin to pay visits to the other nucleus and to exchange places. It is easy to see that the two separate hydrogen atoms thereby become a single stable system H—H. For if we imagine the two nuclei brought very close together, so that they almost form a doubly charged nucleus, we have exactly the configuration of the helium atom, which, as we know, has a stable ground-state. In reality the nuclei repel one another, and hence will not come so close as all that, but the orbits (or the corresponding vibrations) of the two electrons will not differ very much from those of the helium atom. In the lowest state of the helium atom, however, the spins of the two electrons are antiparallel, by the Pauli principle. We have a closed shell. Exactly the same applies to the H—H molecule. If another hydrogen atom comes along, its electron cannot penetrate into the closed shell of the molecule; hence there is no marked interaction, and the H_2 molecule does not react with a third hydrogen atom.

We see from this that the spins take over the parts of the little hooks or valency bonds of the chemist. If two electrons have their spins antiparallel, one valency is saturated. The valency of the atom should accordingly be equal to the number of

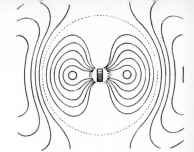

electrons whose spins have no antiparallel partners. This corresponds roughly to the facts of the periodic system. There are, however, apparent exceptions, which require special consideration. The troublesome investigation of details, which must be carried out by physicists and chemists in combination in order to elucidate these difficulties, is by no means complete. Yet there is no possible doubt that the quantum theory is capable of explaining all the properties of atoms and molecules accurately, although the working-out of details may still leave very much to be desired. The riddle of matter is not indeed solved, but is reduced to a deeper problem, which, however, is in many ways simpler, namely, what are the atomic nuclei?

Before we attack this problem, however, we shall consider one of the most remarkable consequences of the quantum theory in somewhat greater detail.

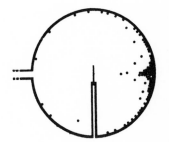

8. *Magnetons.*

We must not forget that in spite of all its successes the quantum theory demands an intellectual sacrifice —renunciation of the complete determinability of position and time for a particle whose momentum and energy are known, and renunciation of the complete prediction of future events. Bounds must be imposed on reason and understanding, because Nature seems to exhibit features which are irrational and unintelligible. Even among concepts which we do seem capable of grasping, there are many which are highly paradoxical. One of the most peculiar is quantization of direction.

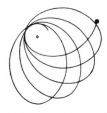

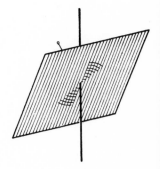

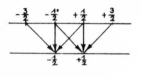

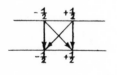

(101)

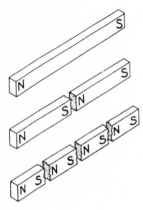

(102)

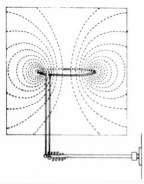

(103)

The idea is that an atom with angular momentum cannot set itself in *every* direction in a magnetic field, but only in a few definite directions which can be counted up. Such an atom is a little magnetized needle. What would happen if a ship's compass were quantized in direction? It would not swing freely to and fro, but would set itself firmly in one direction, and if it were shaken hard it would suddenly jump and point in quite a different direction! In fact, the compass needle must actually behave like this! Fortunately, however, Planck's constant is so small that the sudden changes of direction lie far below any imaginable powers of observation. In the case of atoms, however, it has been found possible to give a direct demonstration of the discontinuity of direction.

Indeed, the Zeeman effect provides indirect evidence of it. Plate III(f) (facing p. 250) shows the splitting of the double D line of sodium in the magnetic field. One line is split into 4 lines, the other into 6. This is easy to understand.

Without the field both states are double; the lower has $l=0$, $s=+\frac{1}{2}$ or $-\frac{1}{2}$, the higher $l=1$, $s=+\frac{1}{2}$ or $-\frac{1}{2}$; the total angular momentum of the latter is therefore either $1+\frac{1}{2}=\frac{3}{2}$ or $1-\frac{1}{2}=\frac{1}{2}$. When the field is applied, the upper state $\frac{3}{2}$ splits up into 4 terms, $m=+\frac{3}{2}, +\frac{1}{2}, -\frac{1}{2}, -\frac{3}{2}$, while the other upper state and the two lower states split into 2 terms, $m=+\frac{1}{2}, -\frac{1}{2}$. From the four upper terms, as (101) shows, there are 6 transitions to the two lower ones; from the two upper terms there are 4 transitions to

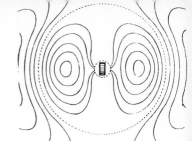

the two lower ones. Hence one D line splits into 6 components, the other into 4.

All Zeeman effects can be explained by simple arguments similar to the above. Here, however, we have only observed the transitions between terms, not these terms themselves, that is, definite states.

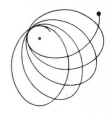

This, too, has been done successfully. We compared the atom to a magnetized needle—an analogy which is much older than the quantum theory. More than a hundred years ago Ampère asserted that there is no magnetism, in the sense in which there is electricity. For if a magnet is broken in pieces, each bit immediately shows a north pole and a south pole (102). The two kinds of magnetism, unlike the two kinds of electricity, cannot be separated. Again, we know from Oersted's experiment (p. 71) that an electric current is surrounded by a magnetic field. If an almost completely closed loop of wire has an electric current passing through it, the magnetic lines of force look very like those of a small magnet whose axis is at right angles to the plane of the loop (103). If the wire is wound into a coil, the resemblance is still closer (104). Such coils are used in all sorts of ways as electromagnets.

Ampère proved that it was not merely a case of resemblance, but of complete identity. He accordingly put forward the hypothesis that in the atoms of magnets and magnetizable substances (such as wrought-iron or nickel) little electric currents are flowing, that is, the atoms are electromagnets. This at once explains the peculiar fact that the magnetic

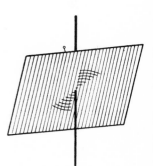

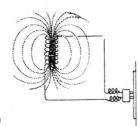

(104)

poles cannot be separated, and is, moreover, in agreement with experimental facts.

Present-day atomic theory has given reality to Ampère's hypothesis. The electrons revolving about the nucleus are circular currents of this kind. Hence every atom is really a little compass needle. These compass needles, however, are peculiar in that they are not to be had in all sizes, but only in one smallest pattern and multiples of it. For the angular momentum is of course a whole number, and it is intuitively evident that the magnetic intensity of the circular current is proportional to it.

The electron itself, however, has an angular momentum, called the spin; it is therefore a magnet, the unit magnet, the so-called *magneton*.

The angular momentum of the electron is of course $\frac{1}{2} \times h$, but according to experiment and theory the magnetic intensity is to be taken without this $\frac{1}{2}$. It is this difference that gives rise to the anomaly of the Zeeman effect. Otherwise, so many of the 4 or 6 Zeeman components of the D lines, say, would coincide that we should have the normal splitting into 3 components.

In addition to the spin magneton, we also have the magnetic effect of the orbital motion, which is always a multiple of the magneton. In the hydrogen atom, and also in the sodium atom, the ground-state has no orbital angular momentum ($l = 0$); here, therefore, there is only the spin magneton. In other atoms with several electrons, however, there are ground-states with orbital angular momentum.

How can the magnetons actually be demonstrated?

(105)

(106)

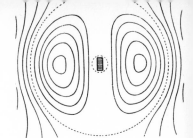

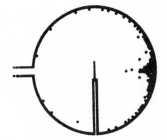

Their total effect is manifested in the large-scale magnetization of bodies. If a needle of wrought-iron is brought near a strong magnet, it becomes magnetic itself. From the intensity of this induced magnetization we could determine the magnetic intensity of a single atom, by dividing by the number of atoms, were it not for the complications arising from the molecular motions (temperature). Only very rough determinations of the size of the magneton can be made in this way.

Stern and Gerlach, however, have succeeded in making accurate measurements of it, using the method of molecular beams.

If a beam of magnetic atoms is produced and allowed to pass through a uniform magnetic field, there is no visible effect on the beam (105). True, all the little magnets tend to set themselves in the direction of the field, but there is no deviating force acting on them. For the field pulls the north pole of the magneton downwards just as strongly as it pulls the south pole upwards. Stern hit on the amazingly bold idea of making the magnetic field non-uniform, so that the forces on the two poles would be slightly different; then the two forces would not balance, and the atom must be deviated as a whole.

The attempt proved successful. One pole of the magnet was made with a sharp knife-edge, the other with a groove (106). Then close to the knife-edge the magnetic field actually varies noticeably in an interval equal to the diameter of an atom, one ten-millionth of a millimetre, and the beam is deflected.

If ordinary mechanics held good, the magnetons

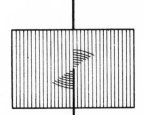

would move into the field in all possible directions. The magnetons lying at right angles to the direction in which the field varies would not be deviated at all, those lying along this direction would be deviated most of all, and between them there would be intermediate deviations. The molecular beam would therefore merely be broadened; a blurred spot would be found on the receiver. Actually, we obtain a number of clearly separated lines; in the case of silver, two, as shown in Plate III(d), facing p. 250; the same for sodium and hydrogen.

This is a direct demonstration of quantization of direction; the magnetons can only take up two positions in the field—along the direction of the field, or in the opposite direction.

From the distance between the two lines we can calculate the size of the magneton; the result agrees excellently with the theoretical calculation.

Very recently Stern has even succeeded in demonstrating the magnetism of the nuclei, which is several thousand times smaller, and Rabi has devised an ingenious instrument by which the angular momenta and magnetic moments of nuclei can be determined with a high degree of accuracy. This brings us to the last stage of our journey through the restless universe, which will take us into the interior of the nucleus.

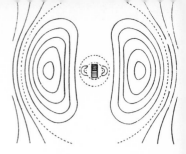

CHAPTER V

Nuclear Physics

1. *Radioactivity.*

OUR journey into the interior of matter is like going down a deep mine. Strata after strata pass by, first the sediments dating from the more recent epochs of the earth's history, rich in fossil animals and plants, then the older strata dating back to times before life existed on the earth at all; finally, if the mine were imagined to go deep enough, we should come to the ultimate magnetic rock, which forms the core or nucleus of the earth.

We, too, in our journey have reached the core or nucleus of the atom. To show the relative magnitudes of the various layers or strata we have passed through, we have made use in (107) (p. 232) of what is known as a scale of orders of magnitude. The centre corresponds to one centimetre; each line is an equal distance from the next, but means a length ten times as small or ten times as great. This does not, it is true, give us an immediate picture of the relative magnitudes, which quite transcend our powers of representation; but with a little trouble an idea of the magnitudes can be derived from this diagram.

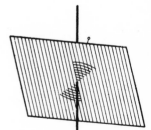

We began our journey far out among the fixed
stars. The nearest of the fixed stars is at a distance
of about 10^{18} cm.; this means that light takes
several years to reach us from there. Our more
immediate home in the universe, the solar system,

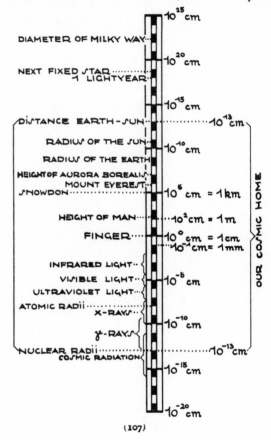

(107)

is thus fairly isolated. The radius of the earth's orbit is about 10^{13} cm.; the radius of the atomic nuclei about 10^{-13} cm. The sun whose radiation makes our life possible, and the nuclei which form the greater part of the matter of which our bodies are built up, are therefore at about equal distances upward and downward on the " order-of-magnitude " scale.

Here it is the lower parts of this scale that we are concerned with. After a journey of 10^{-8} cm. we reach the atoms in their thermal dance. We penetrate into them, and far from finding a state of greater rest, we find still wilder motion. The electrons in the inner shells of a light atom like lithium vibrate about 10^{17} times per second—a colossal number! Let us compare it with long intervals of time and ask ourselves the question, what happened 10^{17} seconds ago? In one year there are $60 \times 60 \times 24 \times 365 = 3 \times 10^7$ seconds, so that 10^{17} seconds come to about 3×10^9 or 3000 million years. This is a longer time than that which has elapsed since the formation of the first solid crust on the earth. With the heavier atoms, the number of times the inner electrons vibrate in a single second is many times greater than the number of seconds since the " creation of the world "!

Now we have reached the nucleus and have hopes of greater rest, firmness, and solidity—but we find none. True, the nuclei are much heavier than the electrons, and as a whole they therefore move correspondingly less rapidly. What goes on in their interior, however, does not promise peace or repose.

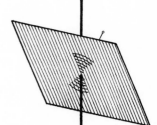

For if we wanted, say, to sit down on a nucleus to rest, it would be advisable to have a good look at it first; otherwise it might explode under us like a shell.

Not all nuclei are explosive, of course; otherwise matter could not continue to exist. The nuclei which have a tendency to explode are naturally rare, for in the course of the history of the universe they have died out. Hence it was not exactly easy to detect them and find out their properties. In fact, this is only possible at all because the explosions occur with immense force, and the fragments shot out betray the presence of explosive nuclei even when only very few of the latter are present. In 1896 Becquerel discovered that minerals containing the element uranium emit a radiation which ionizes gases and blackens a photographic plate. A couple of years later the two Curies succeeded, after a troublesome process of working up many tons of pitchblende (a uranium ore), in separating the radiating constituent; it turned out to be a metal related to barium, and was given the appropriate name of *radium*. The idea that the phenomena were due to atoms exploding was put forward by Rutherford and Soddy in 1903. They showed that radium gives rise to a gas, which they called radium emanation (nowadays it is usually called radon); this again explodes, leaving a solid deposit behind, which disintegrates in its turn, and so on, giving a whole series of *disintegration products*, which we shall discuss below. Two other disintegration series have been discovered, one of which is connected with the

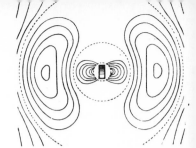

radium series. We must, however, first explain how it is possible to distinguish and isolate these substances, which usually occur in exceedingly minute quantities, invisible and unweighable, and often have but a short life.

This is done by studying their " rays ", that is, the fragments shot out in the explosions; not by distinguishing between the kinds of fragments— they are always the same few kinds of particles —but by counting their number.

2. *The Disintegration Laws.*

We have already used the rays from radioactive substances in our investigations of the atom. We know the methods for determining their specific charge, measuring their intensity in the ionization chamber, counting them, photographing their tracks in the Wilson chamber, and so on (p. 95). We found that there are three kinds of radioactive rays, which are denoted by alpha (α), beta (β), gamma (γ), the first three letters of the Greek alphabet.

Alpha-rays are positively charged ions. Their specific charge is equal to that of the doubly charged helium atom. As helium has only two electrons, the α-particles must be helium nuclei. It was actually found possible to show that when a small glass vessel in which α-particles had been collected was used as a Geissler tube it gave the spectral lines of helium.

The fragments also include fast electrons, called β-rays. Finally, there is an electrically neutral radiation, called γ-rays; these are in every respect identical with very penetrating X-rays.

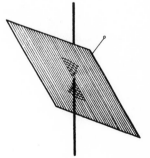

The most important method for analysing radio-active substances consists in observing the decay of one or other of these radiations. The decay curves obtained are usually very complicated; for even if we start with a pure element, it gives rise to fresh elements, which go on disintegrating and emitting other rays. In fact, it may even happen that the radiation, instead of steadily declining, actually rises; this happens when a product of disintegration radiates more strongly than its predecessor.

We then try to separate the substances chemically. Known substances are added to the solution of the radioactive mixture, and we watch whether the source of the radiation, or of part of the radiation, remains in solution or comes down with the pre-cipitate. In this way the Curies showed that the activity of pitchblende remains with the element barium and is only separable from it with great difficulty. The carrier of the radiation, the radium, must therefore be an element resembling barium; as a matter of fact, it fits into the periodic table below the alkaline earth metals magnesium, stron-tium, barium. It was also found that a disintegration product of the gaseous radium emanation can be precipitated along with tellurium, and its successor with lead. These are now called radium A and radium B.

Once these separations are carried out the decay curves become simpler and smoother. This indicates that we are approaching a single substance. Once this is obtained, the *law of decay* is always the same; it is a law which plays a considerable part in science

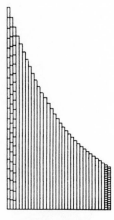

(108)

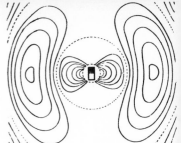

and also in everyday life. Suppose that someone
borrows a sum of money on which he has to pay
yearly interest of 5 per cent. In addition to the
interest, he wishes to pay back 5 per cent of the
borrowed capital each year. Then the debt will
decrease every year by a twentieth of what is still
left (108). The law of radioactive disintegration
is like this. In every small interval of time, the
same fraction of the atoms which are still left disintegrates. No matter what the quantity we start
with, by a perfectly definite time exactly half will
have disintegrated; this is called the *half-value
period* of the substance. For radium it is 1590 years.
Some radioactive substances, however, have much
longer half-value periods; thorium, for example,
has a half-value period of 2×10^{10} years. In others
it is exceedingly brief; a disintegration product of
thorium, for example, has a half-value period of only
10^{-9} second. Periods as large or as small as this,
of course, can only be measured indirectly.

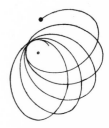

The nature of this *disintegration law* indicates
right away that we are dealing with a *statistical* law.
This has been confirmed in a great variety of ways.
Above all, it was found quite impossible to affect the
disintegration by any means whatever, whether by
high or low temperatures, electric or magnetic fields,
or any other influences. This shows at the same
time that radioactivity cannot reside in the outer
electronic layers, but must be a *nuclear process*.
That the phenomena are of a statistical nature can,
moreover, be proved accurately by experiments on
distribution, such as we are familiar with in the

kinetic theory of gases. We count (with a Geiger counter, say) the number of α-particles shot out during a considerable time, which, however, is short compared to the half-value period. If we divide this time into ten portions, say, we do not by any means obtain the same number of particles in each portion, but their number fluctuates irregularly (109). These fluctuations are found to be in accordance with the laws of probability, which means that the explosions take place wholly at random. No nucleus knows how long it will live—it may explode the next instant, or it may last for many years. Only one thing is certain: if we have a great number of nuclei, we can bet that their decay will *on the average* accord with the law of disintegration asserted above. Radioactivity was the first phenomenon which showed elementary processes exhibiting a purely statistical regularity. We may cherish the opinion that there must ultimately be some inner reason for the fact that one atom lives only a few seconds and its apparently identical neighbour many years; but no one has yet succeeded in putting his finger on the cause. And the new quantum mechanics declares that there is no sense in looking for it.

This is not merely a renunciation, but at the same time a gain. For the quantum mechanics enables us to explain another law of disintegration, which is quite unintelligible from the point of view of ordinary deterministic mechanics.

According to Geiger and Nuttall, there is, in fact, a relation between the length of life of a nucleus and the velocity of the α-particle shot out from it.

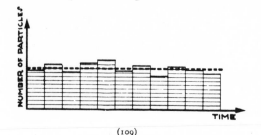

(109)

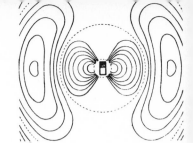

The shorter the half-value period of the nucleus, the faster the α-particle, the relation being shown by the curve in the figure (110). On the basis of ordinary mechanics, it is really impossible to explain the explosion of the nucleus. In ordinary mechanics a

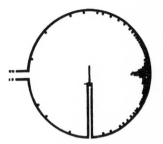

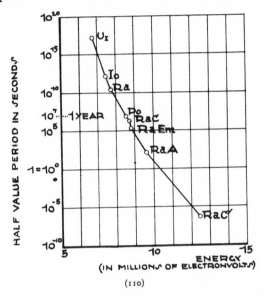

(110)

system of particles is either stable or unstable; in the first case it persists indefinitely, in the second it breaks down at once. To devise a mechanism which would explain the statistical regularity of the explosions would be very difficult.

From the point of view of wave mechanics, however, the problem immediately becomes intelligible, and it brings out the characteristic features of this

theory in a particularly beautiful way. Here a fairy tale may appropriately be told.

Once upon a time there were two boys who rescued a gnome from a serpent in the forest. In gratitude he gave each of them a money-box of a peculiar kind. These money-boxes took the form of earthenware balls, and when they were shaken the agreeable chink of gold pieces was heard. But no opening for putting in or taking out money could be seen. " That's not necessary," said the gnome, " only shake the ball vigorously and the gold will jump out, and I will see to it that the box is never empty." The boys began to shake their money-boxes. But one of them soon got impatient, lost his temper, and smashed the money-box. He found one gold piece inside and that was all. The other boy was a " good boy " and went on shaking his box and enjoying the pleasant chinking sounds. Suddenly a gold piece jumped out, without there being any hole in the earthenware ball. The boy went on shaking and shaking, and from time to time, after longer or shorter intervals, a gold piece came out, till at last he became a rich man. But as he was not only good and patient, but clever too, he pondered over the magic box and soon found out how it worked: the gnome had simply appointed the sound-waves of the chinking gold pieces to be their de Broglie waves. He published this discovery and became a celebrated physicist, just like Gamow, who discovered the same magic in the case of the atomic nuclei.

For what we have just described as happening in

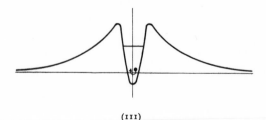

a fairy tale, pieces of solid matter moving through solid walls, does actually happen when nuclei emit α-particles. The force holding the α-particle in the nucleus may be thought of as like a crater with the particle tearing round inside it. The particle has a definite energy, represented in figure (111) by a straight line; the particle can reach that height, but never surpass it. Hence according to classical mechanics it must be a prisoner in the crater to all eternity—like the gold pieces in the money-boxes in the fairy tale. According to wave mechanics, however, it corresponds to a vibration in the crater: the vibration is by no means confined to the interior, however, but penetrates a little way into the wall of the crater and into the space outside (112). Just as the chinking of the gold pieces could be heard outside the money-box, quite feeble de Broglie waves are continually leaving the nucleus. These indicate that there is a certain small probability of suddenly finding the α-particle outside, rushing on with the same velocity as it had in the interior of the nucleus. In a sense it pierces through the wall of the crater, so that we speak of a " tunnel effect ". We could imagine, instead, that it scrambled over the top of the wall; but then the principle of energy would be contradicted during the climb. This is a matter of taste. Neither idea can be tested by experiment; they are just different ways of picturing a process which cannot be concretely realized.

Geiger and Nuttall's law is now fairly obvious. We have only to assume that the height of the crater wall which is to be surmounted, or, in physical

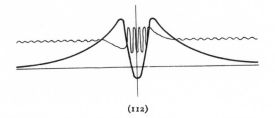

(112)

language, the " potential barrier " or " energy threshold " which is to be passed, is nearly of the same height and thickness for α-particles from various radioactive nuclei. Then it is clear that α-particles of higher energy have only to break through the thinner upper part of the wall, and hence have a better chance of escape than α-particles of lower energy, which have to penetrate the thicker foot of the wall (113).

That this is really so follows from the fact, known from scattering experiments, that Coulomb's law still holds down to a distance of 3×10^{-12} cm. from the nucleus of the heaviest element, uranium; it is then easy to calculate that there the energy of electrical repulsion (that is, the height of the crater) is more than twice that of the emerging α-particle.

We thus have a beautiful application of wave mechanics to nuclear phenomena, even although the crater comparison must be regarded as very rough and sketchy. Before we attempt to refine it, we must learn a great many more facts about the structure of nuclei.

3. Isotopes.

If the disintegration theory is correct, that is, if radioactivity is a question of the formation of fresh nuclei by the splitting-off of fragments (helium nuclei or electrons), the periodic system of the elements is at once enriched by a whole crowd of new elements. The pressing matter now is, do these fit into the periodic system? We were brought up in the belief that practically all the elements were

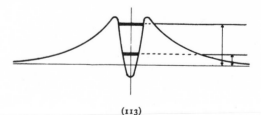

(113)

PLATE I

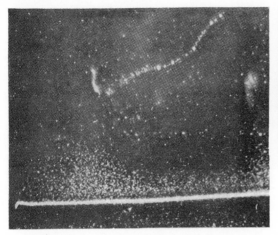

(a) Tracks of an electron and an α-particle in a Wilson chamber

(b) Circular waves

(c) Straight water waves transformed into circular waves on passing through a slit

(e) Diffraction through a narrow slit

(d) Interference of two circular water waves

(f) Diffraction through a wide slit

already known. Now a few dozen more turn up
and have to be fitted in somewhere. It seems impossible, and would indeed have been so, if the older
chemists had been right in holding that the atomic
weight is the important thing, the essential characteristic of a chemical element. Meanwhile, however, physical discoveries have shown that this
idea is false: it is the atomic number, that is, the

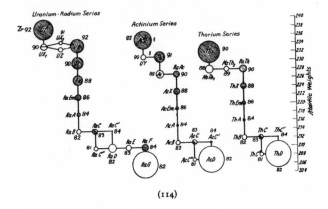

(114)

number giving the nuclear charge, that determines
the number of electrons outside the nucleus, hence
the structure of the swarm of electrons, and hence the
physical and chemical behaviour of the atom. The
nuclear mass is obviously a secondary matter. Why
should it not happen that two nuclei have the same
charge but different masses?

That this does happen was first recognized in the
case of the radioactive disintegration series. There
are three such series, which are shown in (114).

Radium is not the starting-point of a series, but is itself a product of disintegration. The two first series start from uranium, element number 92, the third from thorium, element number 90. The zigzag paths of the series arise from the fact that the emission of an α-particle is indicated by a jump downwards, that of a β-particle by a jump to the right. At the same time, nuclei which emit α-particles are shown shaded, those which emit β-particles being unshaded. The sizes of the circles roughly correspond to the half-value periods, the biggest circles being of course those of the long-lived end-products.

As α-particles are helium nuclei, with charge 2 and mass about 4, an element which emits α-rays must be transformed into one whose atomic number is two units less, that is, one which lies two places farther back in the periodic table. Again, the emission of β-rays raises the nuclear charge by 1; for the electron is negative, so that its loss means an increase of (positive) charge. Hence an element which emits β-rays moves one place forward in the periodic table. This is the " law of radioactive transformation ".

As the chemical nature of many members of the disintegration series can be determined by the precipitation method mentioned on p. 236, the law can be tested experimentally, and has been found to apply throughout. We now see immediately that a particular atomic number occurs not once, but repeatedly, and can even occur more than once in the same series. Thus in the uranium and thorium series

the atomic number of the starting-point, 92, again
occurs as that of the fourth disintegration product.
Ra A, Ac A, and Th A all have the same atomic
number, 84.   More important still, this is true of
all three end-products Ra G, Ac D, and Th D; their
place in the periodic table, 82, however, is already
filled by lead. These four substances, and a few more
(such as Ra D, Ac B, Th B) are therefore of the
same chemical nature—though by no means identical, for they differ in atomic weight. This is shown
by the atomic weight scale on the right. The atomic
weight is actually known, provided only *one* substance in each disintegration series is available in a
quantity sufficient for its atomic weight to be determined directly by chemical methods and the use of
the balance. For then we have only to run through
the series: every emission of an α-particle lowers the
atomic weight by 4, every emission of a β-particle
by $1/1840$, that is, practically not at all.   Now it
was found possible to prove by direct methods that
the three end-products resembling lead *differ* in
atomic weight. It is clear that, if this theory is correct,
lead must occur in all radioactive minerals, owing
to the accumulation of end-products. This is actually
the case. The atomic weights of lead from uranium
ores and of lead from thorium ores can be determined, and different values are actually obtained,
the differences being in the directions indicated by
the theory.

Hence there is no doubt that different nuclei can
exist with the same charge but different masses.
These are called *isotopes* (Greek *isos*, equal; *topos*,

place, that is, elements which occupy the same place in the periodic table).

Now it seemed very probable that the phenomenon of isotopy might also occur with other non-radioactive nuclei. For we know that although the atomic weights exhibit a definite tendency towards whole numbers (for example, Li 6·94, Be 9·02, C 12·00, N 14·008), there are some quite marked deviations from whole numbers among them (for example, Mg 24·32, Cl 35·457, Zn 65·38). Might not the old hypothesis of Prout, which we mentioned in Chapter II (p. 56), be true when applied, not to the whole atoms, but to the atomic nuclei? Might not pure isotopes be merely clumps of protons—perhaps with a few electrons to stick them together—and might not the chemical, non-integral atomic weights arise from the mixing of isotopes? The latter is actually the case. The method by which J. J. Thomson succeeded in proving it is that commonly employed in atomic physics: the production of rays consisting of the particles under investigation, and electromagnetic deviation of these rays.

Beams of positive ions can be produced in evacuated tubes in various ways. As was found by Goldstein, for example, they occur in ordinary Geissler tubes if holes or " canals " are made in the cathode (115). The electrons emitted by the cathode ionize the atoms of gas with which they collide, and the positive ions thus produced fall on to the (negative) cathode, on which, as a rule, they are caught. If, however, an ion happens to land in a perforation, it

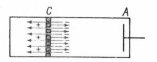

(115)

PLATE II

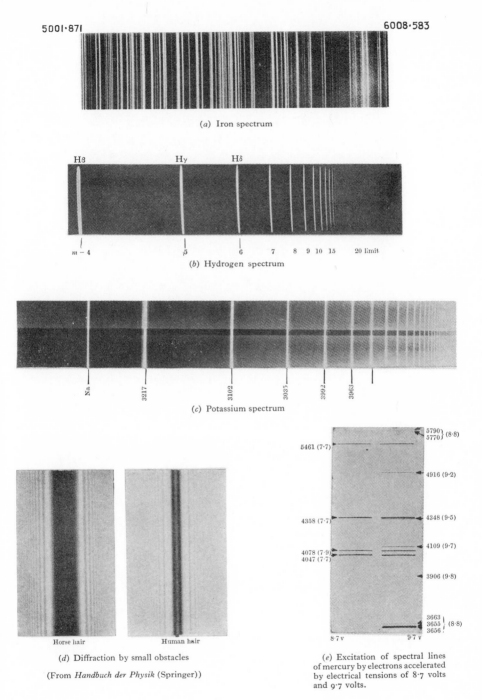

5001·871   6008·583

(a) Iron spectrum

Hβ   Hγ   Hδ

m − 4   5   6   7   8   9   10   15   20 limit

(b) Hydrogen spectrum

Na   3217   3102   3035   3992   3963

(c) Potassium spectrum

Horse hair   Human hair

(d) Diffraction by small obstacles

(From *Handbuch der Physik* (Springer))

5790 }
5770 } (8·8)
5461 (7·7)
4916 (9·2)
4358 (7·7)   4348 (9·5)
4078 (7·9)
4047 (7·7)   4109 (9·7)
3906 (9·8)
3663 }
3655 } (8·8)
3656 }
8·7 v   9·7 v

(e) Excitation of spectral lines
of mercury by electrons accelerated
by electrical tensions of 8·7 volts
and 9·7 volts.

passes through it and out at the other side. This type of *positive rays* is sometimes given the name of *canal rays*.

If the cathode fits the tube tightly, so that no gas can enter the region behind it, a high vacuum can be produced on the side where the positive rays emerge, for very little gas can get through the fine perforations. The positive rays then have quite a long range. In the case of many substances positive rays can also be produced by spreading a thin sheet of the substance over the anode; atoms are then torn out of this, losing their electrons, that is, becoming positive ions. We then have *anode rays*.

These positive rays can be deviated and detected photographically, just like cathode rays. As a rule we obtain a number of different ions, as shown in Plate V(*a*), facing p. 258; each of the parabolic streaks corresponds to one particular kind of ion, the ratio $\frac{\text{charge}}{\text{mass}}$ being the same for different points of the same parabola, the velocity of the ion alone varying from point to point.

In this way J. J. Thomson succeeded in proving that the gas neon, for example, has, in addition to the main parabola, a feebler one corresponding to a mass greater by 2 units.

The method was brought to a high degree of perfection by Aston. His apparatus (116), known as the *mass spectrograph*, is so adjusted that ions of one kind but of differing velocity do not meet the photographic plate along a parabola, but are united at a single point, just as light is at the focus of a

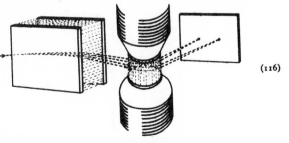

(116)

lens. The sensitiveness of the apparatus is thus greatly increased, the accuracy of the mass determinations actually being considerably higher than that of the chemist.

Plate V(c), facing p. 258, shows a so-called mass spectrum, that is, the spots on the plate which correspond to the different kinds of ions. A glance at it will suffice to convince anyone of the integral nature of the atomic weights; the spots are at regular distances apart, which are all integral multiples of a certain minimum distance. This has of course been confirmed by careful measurements—in which, however, deviations again became evident, which needed fresh arguments to explain them.

At any rate, Prout's hypothesis holds in the following form:

All nuclei have masses which, apart from small deviations, are integral multiples of the mass of the proton. The integer is called the *mass-number*.

As a glance at Table III (p. 249) shows, almost all chemical elements are found to be a mixture of isotopes. This is the reason for the larger deviations of the chemical atomic weights from whole numbers.

Attempts have naturally been made to obtain the separate isotopes pure, and in many cases this has actually been done. For example, the two lithium isotopes of masses 6 and 7 can be separated by electromagnetic deviation of their positive ions, at least to such an extent that the extremely thin deposits obtained can be used for all manner of other experiments (such as bombardment of these deposits with other particles, and so on). In the

TABLE III

TABLE OF ISOTOPES

The isotopes are arranged in each case in the order of frequency of occurrence; the radioactive isotopes are indicated by an asterisk. Radioactive isotopes produced artificially are not included.

Element	Z	Isotopes	Element	Z	Isotopes
H	1	1, 2, 3	Sn	50	120, 118, 116, 119, 117, 124, 122, 121, 112, 114, 115
He	2	4, 3			
Li	3	7, 6			
Be	4	9	Sb	51	121, 123
B	5	11, 10	Te	52	130, 128, 126, 125, 124, 122, 123, 127?
C	6	12, 13			
N	7	14, 15	I	53	127
O	8	16, 18, 17	Xe	54	129, 132, 131, 134, 136, 130, 128, 124, 126
F	9	19			
Ne	10	20, 22, 21	Cs	55	133
Na	11	23	Ba	56	138, 135, 136, 137
Mg	12	24, 25, 26	La	57	139
Al	13	27	Ce	58	140, 142
Si	14	28, 29, 30	Pr	59	141
P	15	31	Nd	60	146, 144, 142, 145, 143
S	16	32, 34, 33	Sm	62	144, 147, 148, 149, 150, 152, 154
Cl	17	35, 37			
A	18	40, 36, 38	Eu	63	151, 153
K	19	39, 41*	Gd	64	155, 156, 157, 158, 160
Ca	20	40, 44, 42, 43	Tb	65	159
Sc	21	45	Dy	66	161, 162, 163, 164
Ti	22	48, 50, 46, 47, 49	Ho	67	165
V	23	51	Er	68	166, 167, 168, 170
Cr	24	52, 53, 50, 54	Tm	69	169
Mn	25	55	Yb	70	171, 172, 173, 174, 176
Fe	26	56, 54, 57	Lu	71	175
Co	27	59	Hf	72	176, 178, 180, 177, 179
Ni	28	58, 60, 62, 61, 56(?), 64(?)	Ta	73	181
Cu	29	63, 65	W	74	184, 186, 182, 183
Zn	30	64, 66, 68, 67, 70	Re	75	187, 185
Ga	31	69, 71	Os	76	192, 190, 189, 188, 186, 187
Ge	32	74, 72, 70, 73, 76			
As	33	75	Hg	80	202, 200, 199, 201, 198, 204, 196
Se	34	80, 78, 76, 82, 77, 74			
Br	35	79, 81	Tl	81	205, 203, 207*, 208*, 210*
Kr	36	84, 86, 82, 83, 80, 78			
Rb	37	85, 87*	Pb	82	208, 206, 207, 204, 203?, 205?, 209?, 210*, 211*, 212*, 214*
Sr	38	88, 86, 87			
Y	39	89			
Zr	40	90, 94, 92, 96, 91	Bi	83	209, 210*, 211*, 212*, 214*
Cb	41	93			
Mo	42	98, 96, 95, 92, 94, 100, 97	Po	84	210*, 211*, 212*, 214*, 215*, 216*, 218*
Ru	44	102, 101, 104, 100, 99, 96, 98?	Rn	86	222*, 219*, 220*
			Ra	88	226*, 223*, 224*, 228*
Rh	45	103	Ac	89	227*, 228*
Ag	47	107, 109	Th	90	232*, 227*, 228*, 230*, 234*
Cd	48	114, 112, 110, 113, 111, 116, 106, 108, 115			
			Pa	91	231*, 234*
In	49	115, 113	U	92	238*, 234*

case of mercury the separation has been carried out by making use of the fact that the heavier isotope evaporates rather more slowly than the lighter. As was shown by Hertz, many gases can be separated by making them pass through the pores of earthenware cylinders; the lighter atoms pass through more rapidly than the heavier, and if the process is repeated sufficiently often, the separation of the isotopic gases is eventually complete.

This, of course, is extremely important. For if we wish to investigate the nucleus more closely, we must be sure that we have a definite kind of nucleus to work with. But two isotopic nuclei, such as the two boron nuclei with masses 10 and 11 mentioned above, differ from one another just as much as they do from the nuclei of the following element, carbon, with masses 12 and 13. The fact that owing to their equal charges they are surrounded by identical swarms of electrons, and hence are chemically indistinguishable, is of quite minor importance so far as the nuclei are concerned. From the point of view of nuclear physics, the electronic shells are masks which disguise the true appearance of the nuclei—simulating identity where no such thing exists.

Conversely, there are also masks which overemphasize a difference. For example, chlorine has an isotope which has the same mass, 39, as the principal isotope of potassium. Such nuclei are said to be *isobaric*, that is, of equal weight (Greek *barys*, heavy). They contain the same number of protons, but differing numbers of electrons. Table III (p. 249) shows a number of such cases.

Plate III

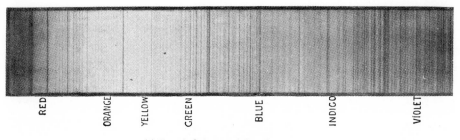

RED ORANGE YELLOW GREEN BLUE INDIGO VIOLET

(*a*) Fraunhofer's map of the solar spectrum

(*By courtesy of the Munich Academy of Science*)

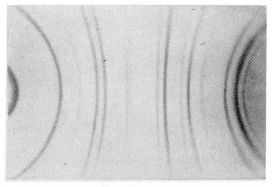

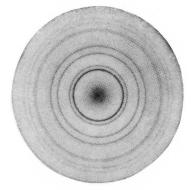

(*b*) X-ray diffraction rings

(From *Physikalische-Zeitschrift* (Hirzel, Leipzig)

(*c*) Electronic diffraction rings

(After H. Mark and R. Wierl)

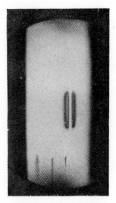

(*d*) Magnetic splitting up of a lithium beam by the method of Stern and Gerlach.

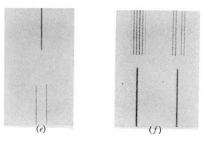

(*e*) (*f*)

(*e*) Normal Zeeman effect
(*f*) Zeeman effect in sodium D lines

We now have the material necessary to enable us
to penetrate farther into the interior of matter:
unmixed nuclei, all of which are clearly made up of
protons. Now it does not seem too difficult to find
out how the nuclei are built up. The next lightest
nucleus after the proton is that of helium. Its mass
is 4 and its charge 2, so it presumably consists of
four protons and two electrons; these are shown in
figure (117) in rather a fanciful arrangement. The
lighter lithium isotope with mass 6 and charge 3
will consist of six protons and three electrons, and
so on (118).

(117)

We can even tell what the energy of binding of
these nuclear systems will be. We have only to
recall that the masses of the nuclei exhibit small
deviations from integral values. We know that a
heaping-up of mass implies a storing-up of energy.
The small *mass-defects*, as the deviations from exact
multiples of the mass of a proton are called, show
how much energy is liberated in the building-up of
the nuclei. An example will make this clear. The
helium nucleus possesses no isotope in any noticeable quantity. Yet its atomic weight 4·002 differs
by quite an appreciable amount from four times the
weight of a proton, $4 \times 1·0078 = 4·0312$; in fact the
difference 0·029 is relatively greater (per proton) than
for most other nuclei. This means that a particularly
large amount of energy is required to split up the
helium nucleus: its binding is very strong.

(118)

The numerical result at once explains why in
radioactive disintegration α-particles, that is, helium
nuclei, are shot out whole.

(B 969)

To obtain the binding energy in ordinary mechanical units, we have to multiply the mass-defect 0·030 by the square of the velocity of light $(3 \times 10^{10}) \times (3 \times 10^{10}) = 9 \times 10^{20}$. We then obtain an amount of energy which, compared with the energy of ordinary chemical processes, is enormous; it is many million times greater than the heat of combustion of coal—both being calculated for the same number of atoms.

The mass-defects of all nuclei exhibit definite regularities. They increase as we pass to elements of higher nuclear charge; from the middle of the periodic table onwards the rise becomes slower, falling off rapidly as we approach the radioactive nuclei. This is not surprising; it merely indicates that these nuclei are unstable and have a tendency to explode.

What sort of forces are these which are developed in the nuclei? And why do the particles crowd together into such a narrow space under their influence? Why are the protons always in excess, so that the nuclei are always positive? Could negative nuclei not exist? Or at least nucleus-like particles in which the number of protons is equal to the number of electrons, " neutrons "? (These could not, of course, form the nuclei of atoms, for they could not attach external electrons to themselves.)

These problems long formed matter for speculation, since nothing definite was known—until a succession of fresh discoveries threw some light on them.

4. *The Deuteron.*

Hydrogen and oxygen play a special part in practical physics. Their compound, water, is used as a standard substance in many measurements. For example, when the metric system of units was set up, the unit of mass (the gramme) was originally defined as the mass of a cubic centimetre of water at a definite temperature. Again, the centigrade scale of temperature is associated with the boiling-point and the freezing-point of water. Similarly in many other cases. Oxygen serves as the standard of atomic weight, and the hydrogen nucleus, the proton, is the unit of nuclear structure. All this is based on the idea that pure hydrogen, pure oxygen, and pure water are well-defined substances.

The first of the discoveries which we have to consider plays havoc with this assumption: for both hydrogen and oxygen are found to be mixtures of isotopes. Although in both cases the quantity of the one isotope is enormously large compared with that of the other, yet in the case of hydrogen, at least, we are faced with the serious fact that the rarer isotope is just twice as heavy as the ordinary one. In all other cases the difference between the weights of the isotopes is fairly trifling compared with their own weights. For this reason they are difficult to separate and play no great part in the everyday life of the physicist or chemist. *Heavy hydrogen*, however, is really quite a different substance from ordinary hydrogen, and hence has been

given a special name. The nucleus with charge 1
and mass 2 is called a *deuteron* (Greek *proton*, the
first; *deuteron*, the second), and the corresponding
element (heavy hydrogen) is often called *deuterium*,
its symbol being D.

The history of the discovery of the deuteron ex-
emplifies how minute deviations of measurements
from the values expected on theoretical grounds
lead to the conviction that a new body exists, and to
its actual discovery. The astonishing thing is the
firm faith which experimenters have in the exactitude
of their determinations. This sort of occurrence,
however, is nothing new. The planet Neptune was
discovered because small deviations in the orbits of
the other planets could not be explained unless by
the assumption that some unknown body was per-
turbing them; its orbit was predicted from the
deviations, and the planet was actually found in the
very spot where it was expected. The discovery of
the rare gases of the atmosphere forms another example. Here it was the minute discrepancy between
the densities of " atmospheric nitrogen " (the gas
left after oxygen is removed from air) and nitrogen
obtained from one of its compounds that led to the
suspicion that an unknown constituent (argon) was
present, and to the proof of its existence.

One method of discovering isotopes is by spectroscopy. We go back to the result, deduced from Bohr's
theory, that the positions of the spectral lines of
atoms depend to a slight extent on the nuclear mass;
we have seen that the spectrum of the singly-charged
helium ion could be distinguished from that of the

PLATE IV

⟶ λ

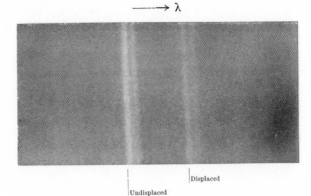

| |
|Undisplaced

|Displaced

(a) Compton effect, Zn Kα

(In *Handbuch der Physik*)

Ag
↓

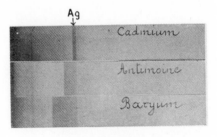

Cadmium

Antimoine

Baryum

(b) The K absorption edge for cadmium,
antimony, and barium

(From Andrade, *The Structure of the Atom*
(Bell & Sons, Ltd.))

hydrogen atom (p. 179). Exactly the same must hold
for two isotopes; their spectral lines will be arranged
in the same way, but will be slightly displaced relative to one another, owing to the difference of
nuclear mass. In this way it was first discovered
that ordinary oxygen, $O = 16$, has two isotopes 17
and 18, though these are present only in extraordinarily small quantities. This discovery led indirectly to the discovery of the hydrogen isotope.
A new determination of the atomic weight of hydrogen relative to the principal oxygen isotope 16 was
made, in which the other two isotopes 17 and 18
were taken into account. The mass of the hydrogen
nucleus had also been determined by Aston's mass
spectrograph, and the two values did not exactly
agree. The difference amounted only to one in five
thousand, but its existence was sufficiently established to give rise to a search for a hydrogen isotope
of mass 2. Urey succeeded in detecting this isotope
spectroscopically. Soon it was possible to obtain it
in an almost pure state by the electrolysis of water.
Molecular hydrogen really consists of three gases,
namely, ordinary H_2, HD_1, and D_2; the latter is of
course only present in very minute amount. Water
also consists of three different sorts of molecules,
H_2O, HDO, and D_2O. In electrolysis the lighter
H_2 is liberated five or six times as rapidly as the
other kinds of hydrogen; hence the heavy isotope
accumulates in the water left behind. By electro-
lysing the residue over and over again, almost pure
heavy water (D_2O) is thus obtained. To-day heavy
water can actually be bought—at least if one has

plenty of money, for the process by which it is manufactured is expensive.

The molecular weight of heavy water is $2 \times 2 + 16 = 20$, as compared with $2 \times 1 + 16 = 18$ for ordinary water; that is, the difference is as much as 10 per cent! The exact mass of the D nucleus is 2·0136; deducting this from the mass of two protons, $2 \times 1·0078 = 2·0156$, we obtain the mass-defect or energy of formation of the deuteron, namely, 0·0020. This is very much smaller than the energy of formation of the helium nucleus, which is 0·030. The properties of " heavy water " (D_2O) differ very considerably from those of ordinary water (H_2O); its freezing-point is 3·8° and its boiling-point 1·4° higher, and its density actually 11 per cent greater!

In all compounds of hydrogen the H atoms can be replaced by D atoms. Hence quite a new branch of chemistry has arisen, which is even of importance in biology.

Fortunately, the discovery of deuterium does not mean an overwhelming catastrophe so far as the certainty of standard physical units is concerned. In any case, we are long past the stage of defining the gramme or unit of mass by means of water. Instead, there is a piece of platin-iridium metal, kept in Paris, which by international agreement is taken to represent the unit 1 kilogram (1000 grammes). In other cases, such as the temperature scale, the effect of the ·minute quantity of heavy water is so trifling that it can be ignored. From the philosophical point of view, however, it is very interesting to see how determinations which at one time were regarded as the firm foundations of physics have subsequently become untenable owing to the refinement of observations.

For the physicist the importance of the discovery of the deuteron lies chiefly in the fact that it is certainly the simplest composite nucleus, whose behaviour when bombarded itself, or when used as a projectile, is very instructive.

5. *The Neutron.*

The idea that the atoms of electricity, the electron and the proton, are the ultimate units out of which matter is built up was a simple and beautiful one. But, alas, it is wrong. There are other particles as well which have an equal right to the title of ultimate atoms.

In the first place, it was found that *neutrons* (p. 252) do actually exist. Their discovery is closely bound up with another discovery, namely, that nuclei can be excited and made to emit light, just as atoms can. This had been suspected for a long time. The γ-rays, of the same nature as light, which are emitted by radioactive substances, can be explained in the following way. When a nucleus explodes, the nuclear residue does not remain in the ground-state, but in an excited state, subsequently jumping back to the ground-state and emitting a proton. Thus the γ-rays indicate the existence of energy-levels in such nuclei as form the end-product of a nuclear explosion. Do these energy-levels not exist in ordinary nuclei also?

Bothe and Becker did actually succeed in causing light nuclei, lithium and beryllium, to emit γ-rays, by bombarding them with α-rays. This is an exact analogy to the excitation of ordinary atoms so that they emit light, by bombarding them with electrons.

As these γ-rays pass through matter they have an ionizing effect, knocking electrons out of the atoms. This process can be traced out in the Wilson chamber; along the γ-ray track we find tracks of ions, which of course always occur in pairs, one positive and one negative.

The study of this phenomenon led Joliot and his wife Irène, daughter of M. and Mme Curie, the discoverers of radium, to make a remarkable discovery. For they found that the secondary radiation emitted by beryllium when it is bombarded by α-rays knocks particles out of paraffin-wax and other substances containing hydrogen (119); and these particles, by their ionizing properties and their tracks in the Wilson chamber, certainly could not be electrons. It turned out that they are protons, and that even the heavy nuclei of helium and nitrogen (Plate V(d), facing p. 258) can be set in motion by the secondary radiation from beryllium. It is unthinkable that the γ-rays should have such an effect; their light-quantum or photon is so light in comparison with the proton that it could no more impart a large velocity to the latter than a tennis-ball could do to a motor-car which it happened to hit.

Chadwick recognized that it was *neutrons* that are liberated from beryllium on bombardment with α-rays. For the point where the proton becomes visible

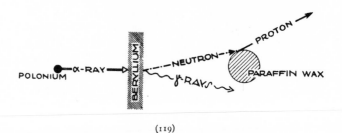

PLATE V

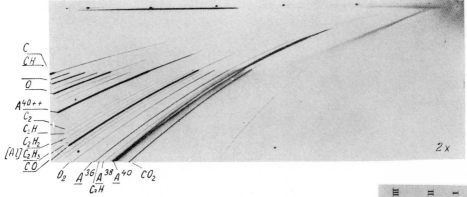

(a) Positive ray parabolas. (After Zeeman & de Gier.)

(By courtesy of the Royal Academy of Amsterdam)

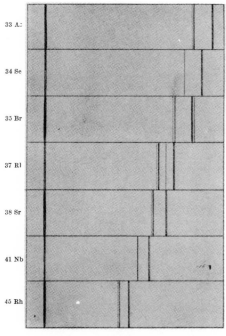

(b) K-series for elements between arsenic
(Z = 33) and rhodium (Z = 45)

(d)

Atom of nitrogen recoiling
after collision with neutron

(From Proceedings of the
Royal Society)

(c) Mass spectrum

From Aston's Isotopes by
courtesy of the publishers, Edward Arnold & Co.

in the form of a cloud-track is not connected with the beryllium by another cloud-track (see Plate V(d)); the new radiation has no effect on the external electrons of the atoms, and therefore cannot consist of anything but uncharged particles. These must have a mass about equal to that of the proton, and if they collide centrally with a proton, they set it in motion. It is even possible to determine the mass of the neutron very accurately, by comparing the results of its collisions with various nuclei, such as those of hydrogen and nitrogen. The result is that the neutron is found to have very nearly the same mass as the proton.

These neutrons behave quite differently from all kinds of rays previously known, whether of light or of charged particles. When the latter pass through matter, the process of retardation and finally of absorption is chiefly a matter of their energy being given up to the external electrons of the substance. As the number of external electrons is approximately proportional to the nuclear mass, the intensities of absorption of different substances are approximately proportional to their masses. Thus a layer of lead has a much more powerful retarding effect than an equally heavy layer of aluminium or of paraffin-wax. In the case of neutrons it is quite different. Neutrons pay no attention to the external electrons, but are retarded only by direct collisions with nuclei. As the latter are almost all about the same size, it is a question of the number of nuclei per cubic centimetre. Here, however, the lighter substance has the advantage. For 1 gramme of hydrogen contains 16 times as many nuclei as 1 gramme of oxygen, i.e. has 16 times as great a retarding effect on neutrons. Neutrons penetrate through thick layers of lead, which are a sure protection against all other kinds of rays, and are held up by thin layers of light substances, which are no obstacle at all to α-rays. Hydrogen has a peculiar effect, in that it does not stop the neutrons completely, but merely retards them; quite slow neutrons can be produced with the help of a layer of paraffin-wax or water. This is obviously due to the fact that the neutrons have about the same mass as the hydrogen nuclei and hence in colliding with the latter give up about half their energy.

What is a neutron like, then? It may be imagined as being a very close combination of an electron and a proton, in which the binding is much closer and stronger than the binding between the electron and the proton in the hydrogen atom. Why two such types of combination should exist, one firm and one loose, is quite unknown.—Yet another new and important discovery followed, which opened up new possibilities of explanation.

6. *The Positron.*

None of the theories of electricity hitherto put forward gives the least clue to any explanation of the fact that positive electricity and negative electricity do not occur in exactly the same way, but in the form of the proton and the electron, with their widely differing masses. It had long been conjectured that positive electrons (of small mass) and

negative protons (of large mass) could also exist;
but it was only the quantum mechanics in its most
refined form, due to Dirac, that led to the definite
opinion that this must be so. The *positive electrons* or
positrons were actually discovered from observations
of a wonderful phenomenon known as the *cosmic
rays*. These in themselves are of very great interest,
and we shall accordingly give a brief account of
them here.

Even in empty space there is no rest. Everywhere
light-waves coming from the luminous stars are
in continual oscillation. Here and there an atom is
found wandering about by itself; their density in
interstellar space is estimated at about 1 atom per
cubic centimetre. Further, the sun is continually
shooting out very fast electrons; these give rise to
the aurora (" northern lights "). For we know that
an electron is deviated by a magnetic field, in a cork-
screw-like path, the coils of which become closer,
the stronger the field and the slower the electron.
Now it is well known that the earth is a magnet.
The electrons shot out by the sun come into the
earth's magnetic field and are deflected into spiral
paths winding round the magnetic poles of the
earth. Størmer has worked out these paths and has
shown that no electrons can reach the equatorial
regions, but that they must accumulate in high
northern and southern latitudes. The phenomenon
can even be imitated experimentally by letting cathode
rays fall on a small magnetized iron sphere, as
shown in Plate VI(*a*), facing p. 262.

These electrons are also responsible for the existence of a layer of ionized gases very high up in the atmosphere, the so-called Kennelly-Heaviside layer, which, strange to say, is of great importance for us. For it is electrically conducting and acts like a mirror for electric waves, just as a metal mirror does for light. This is why our radio transmitters have such a wide range, in spite of the curvature of the earth; the waves cannot get out into space, but are thrown back, once, twice, or many times, and thus reach the surface of the earth at a great distance, so that we can hear music from beyond the Ocean (120).

The electrons from the sun, however, do not actually reach the surface of the earth; but there are other missiles flying through space with colossal energies, some of which do reach us. It is about twenty-five years since this barrage was first noticed on the earth. The ionization chambers used to detect radiations in the laboratory never show absolutely *no* current; counting apparatus, too, always gives a few deflections. This is partly due to the fact that everywhere in the earth there are traces of radioactive substances, which occasionally send a particle through the apparatus. But even when the apparatus is shielded as far as possible from terrestrial effects by thick sheets of lead, there always remains some radiation which cannot be got rid of.

Hess was the first to take an ionization chamber up with him in a balloon; he found that the radiation increased as he went up. Later these experiments were extended to great heights by means of aeroplanes and balloons with or without observers, as in Piccard's celebrated stratosphere flight. The

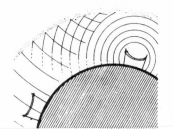

(120)

PLATE VI

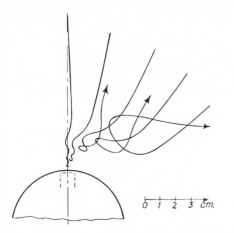

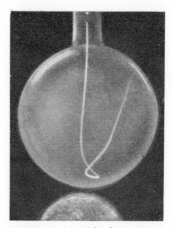

(a) Deflection of cathode rays by magnetized sphere (after Brüche), showing origin of aurora

(From *Physikalische Zeitschrift* (Hirzel))

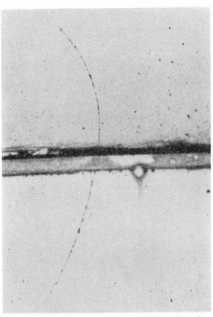

(b) Shower of cosmic rays, showing electrons and positrons

(From *Proceedings of the Royal Society*, 1933)

(c) A 63 million volt positron passing through a 6 mm. lead plate and emerging as a 23 million volt positron.

(*Handbuch der Physik* (Julius Springer))

greatest heights reached with balloons carrying self-registering apparatus are about 30 kilometres. These observations clearly indicate that the radiation falls on the earth from outside. It cannot originate in the highest layers of the earth's atmosphere, say, as a result of the electrical tensions which are discharged in the lower layers of the atmosphere by thunderstorms. For it was proved that the radiation decreases slightly towards the equator; the natural explanation of this is that the radiation consists of charged particles, which come from outer space into the magnetic field of the earth and are deviated by it, just like the electrons which cause the aurora, but to a much smaller extent, owing to their greater velocity. The radiation does not come from the sun, or from the Milky Way, or from any other special direction. It seems to fill all space. Its energy must be immensely great; for it can be detected at the bottom of deep lakes 500 metres below the surface of the water.

At last experimenters succeeded in making the particles of which the radiation consists visible in the Wilson chamber. The whole chamber was put between the poles of a strong magnet, and curved tracks were then seen (Plate VI(*b*), facing p. 262). The energy can be calculated from the curvature, and the particles are found to have energies equal to those of electrons which have been subjected to an electrical tension of several hundred or even thousand million volts.

But these rays, which are observed here on the surface of the earth, are certainly not the original *cosmic rays* alone, but a mixture of these with secondary rays produced by the cosmic rays as they pass through the matter in the atmosphere or apparatus.

It was these rays originating mysteriously somewhere in space that drew attention to the existence of positive electrons. Anderson was the first to notice that often the same plate will exhibit tracks of opposite curvatures, which appear to come from the same point of the wall of the chamber. He was also the first to express the view that they were due to positive electrons. They could not be proton-tracks, for they looked exactly like those of ordinary electrons. The possibility that the tracks curving the wrong way were due to ordinary negative electrons moving backwards seemed improbable, in view of the common starting-point of the tracks; whole showers of particles were found (Plate VI(*b*)), which must obviously arise from a sort of explosion in the nuclei of the atoms of metal in the apparatus when struck by a cosmic particle. At last it was proved directly that the suggestion that the paths are described backwards will not do. A lead plate was placed in the chamber, and particles were occasionally observed to pass through the plate (Plate VI(*c*). Where the path is more strongly curved, the particle is slower; the lead plate can, of course, only *retard* the particles, so the direction of their motion is definitely indicated. The results led perforce to the conclusion that positive electrons occur in the cosmic radiation.

Soon it was also found possible to produce positive electrons in other ways. When light elements are bombarded with γ-rays, electron-pairs are observed to appear in the Wilson chamber, a positive electron and a negative electron shooting out from the same place.

These and similar observations raise a host of questions: why do the positrons occur so rarely in the universe? Do they lie hid in the nuclei? Are they liberated from there by light? Why is a positron always accompanied by a negative electron?

The answers to the last two questions were given by the theory of Dirac mentioned on p. 261, even before the experiments had led to them being asked; and the results have now been confirmed by new direct experiments. A bold assertion is made, and yet one that is consistent with the line which physics has followed from the beginning: matter does not persist from eternity to eternity, but can be created or destroyed. A positive electron and a negative electron may annihilate one another, their energy flying off in the form of light; but they can also be born, with the annihilation of light-energy.

Once the equivalence of mass and energy had been recognized, the possibility that material particles, electrons in particular, can be created or destroyed was often thought of. But now the phenomenon is made visible to our eyes. For in the Wilson chamber we actually see the birth of an electron-pair (Plate VII(a), facing p. 266). The reverse process, the collision and vanishing of a positive electron and a negative electron, has been demonstrated with equal certainty.

The state of affairs is therefore as follows: each electron seeks for a partner of the opposite kind and rushes to unite with it. In this wild wedlock the parents disappear and a pair of twin photons are born. But not all electrons find a partner. In our part of the universe there is a superfluity of the negative kind. Why? We have no idea. In other parts of the universe, perhaps, the reverse may be true.

Perhaps negative protons also exist—no one has succeeded in finding them yet. And perhaps there are regions in the universe where these are in excess. There positive electrons would circulate round negative nuclei. Matter of that kind, however, would not greatly differ from our matter: the inhabitants of such regions would observe exactly the same physical laws. Light might bring us a message that everything there was electrically reversed. But the message would probably be too faint and too indistinct for us to decipher.

7. *Nuclear Transformations.*

We have firmly established the existence of four elementary particles: two light ones, the electron and the positron, and two heavy ones, the proton and the neutron. Probably these are too many. For it is likely that combination of

$$\left.\begin{matrix} \text{a proton and an electron} \\ \text{a neutron and a positron} \end{matrix}\right\} \text{will give} \left\{\begin{matrix} \text{a neutron,} \\ \text{a proton.} \end{matrix}\right.$$

Either neutron or proton must be composite. This leads to the general question of the structure of

PLATE VII

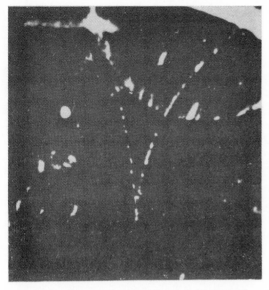

(a) A pair of electrons (positive and negative) generated by a collision between a photon and a gas molecule

(After Curie & Joliot)

(c) Disintegration of lithium nucleus by a proton (two helium nuclei formed)

(b) Disintegration of a nitrogen nucleus by an α-particle; a proton of very long range is emitted

(From *Proceedings of the Royal Society*)

composite nuclei.   But what anatomist's knife is sharp enough to cut asunder bonds which are millions of times stronger than the strongest chemical bonds?

Here is the old problem of the alchemists in a new dress: the actual transmutation of the elements. Now, however, the motive is not the lust for gold, cloaked by the mystery of magic arts, but the scientist's pure curiosity.   For it is clear from the outset that we may not expect wealth too.   It is a question of shooting at and hitting atomic nuclei, and we know how small these targets are: their radius is one hundred-thousandth (10^{-5}) that of the atom, so that the area of the target is $10^{-5} \times 10^{-5}$, that is, 10^{-10} (one ten-thousand-millionth) of the cross-section of the atom.   It is only by the utmost refinement of experimental technique that we can succeed in occasionally scoring a bull's-eye and observing the results.   Rutherford was the first to transmute atoms, by bombarding nitrogen atoms with α-rays.   In the Wilson chamber containing nitrogen, tracks of α-particles are occasionally seen which come abruptly to an end; in their place appears a new track of greater range (Plate VII(b), facing p. 266).   The same result was soon obtained with many other elements.   Deviation experiments showed that the new particles are protons.   At first the phrase " nuclear disintegration " was used, but this expression is misleading.   Clearly it is a case of a *nuclear transformation*, which is actually rather a process of building-up; for the heavy α-particle is retained and the light proton flung out.

It is now customary to express these nuclear transformations symbolically in the same way as chemical reactions. To the symbol of the nucleus we add the mass as a left-hand suffix and the charge (the atomic number) as a right-hand index; for example, we express the nitrogen nucleus by $_7N^{14}$. Then Rutherford's original nuclear reaction is obviously

$$_7N^{14} + _2He^4 \rightarrow _8O^{17} + _1H^1;$$

that is, the α-particle, which is a helium nucleus $_2He^4$, combines with the nitrogen nucleus to form a nucleus of the oxygen isotope $_8O^{17}$, a proton $_1H^1$ being split off. Any other nuclear reaction can be expressed in exactly the same way.

These experiments confirm the idea that the nuclei are built up of protons; for protons are always shot out.

The next step, of course, was to use not only the α-rays provided by Nature, but also *artificial α-rays*. This raised the technical problem of constructing tubes which would give positive rays of far greater energy than any previously used. In addition, huge machines had to be made to give the gigantic electrical tensions necessary, and insulators had to be found which could stand up to these tensions. Many laboratories all over the world worked at the problem, aiming at tensions of more than a million volts. Some were even bold enough to harness lightning; a station was built on Monte Generoso to utilize the immense tensions which occur during thunderstorms.

Yet the first great success was scored by Cockcroft and Walton, using comparatively insignificant apparatus. They produced beams of protons of "only" 120,000 volts, and with these they managed to disintegrate the lithium nucleus, according to the formula

$$_3\text{Li}^7 + {}_1\text{H}^1 \rightarrow {}_2\text{He}^4 + {}_2\text{He}^4.$$

By catching a proton the lithium nucleus acquires the charge 4 and the mass 8; it then breaks down into two helium nuclei, which are, of course, particularly stable structures. In the Wilson chamber it is actually possible to see two helium nuclei flying off in opposite directions as "artificial α-rays" (Plate VII(c), facing p. 266).

Similar reactions can be brought about with many other nuclei. It is usually possible to guess what the nature of the fragments will be; then it is necessary to test whether the energy transformation during the process agrees with the guess. For the mass-defects of the nuclei are known, and hence their binding energies, and the kinetic energies of the projectile and of the resulting fragments can be measured.

Other projectiles can be used besides protons. The deuteron $_1\text{D}^2$ has been found to be particularly effective. If it is shot into heavy water or other substances which themselves contain $_1\text{D}^2$ atoms, reactions are observed which are very probably to be interpreted as follows:

$$_1\text{D}^2 + {}_1\text{D}^2 \rightarrow {}_1\text{H}^3 + {}_1\text{H}^1$$
and $$_1\text{D}^2 + {}_1\text{D}^2 \rightarrow {}_2\text{He}^3 + {}_0\text{n}^1,$$

where $_0\text{n}^1$ denotes the neutron. Thus new isotopes of hydrogen and of helium are produced, which are isobaric with weight 3. The existence of the hydrogen isotope $_1H^3$, which was naturally given the name *triton*, was soon afterwards confirmed by the mass spectrograph.

An important discovery was made by Irène Curie and Joliot. They bombarded aluminium with α-rays and found that after the bombardment had ceased the aluminium itself emitted rays: it had become radioactive, emitting positive electrons. Two reactions are involved; the first,

$$_{13}Al^{27} + {}_2He^4 \rightarrow {}_{15}P^{30} + {}_0n^1,$$

consists in the transformation of the aluminium nucleus into a phosphorus nucleus by the capture of an α-particle, a neutron being emitted. This phosphorus nucleus, however, is not the ordinary stable one, for though it has the same charge 15, its mass is 30 instead of 31. Thus a new isotope of phosphorus has been produced, which is unstable; it explodes in accordance with the formula

$$_{15}P^{30} \rightarrow {}_{14}Si^{30} + \text{positron},$$

becoming transformed into the stable isotope of silicon, with the emission of a positron. According to Fermi, a particularly large yield is obtained by bombardment with neutrons, especially if these are slowed down by passage through a layer of paraffin-wax.

The half-value period of " radio-phosphorus " amounts to only three minutes. In other cases shorter or longer half-value periods have been found, ranging up to about half an hour. These times are long enough for chemical reactions to be performed with the substances and tests made to find out with what elements they are precipitated. In this way the correctness of the interpretation of many nuclear reactions has been confirmed.

Radioactive isotopes of all the elements of the periodic system have been found, and the system itself has been extended beyond uranium 92, by the discovery of several "transuranium" elements, amongst them the notorious nuclei neptunium, Np^{93}, and plutonium, Pu^{94}, which disintegrate by "fission" (see Note to Table I, p. 51, and Postscript).

A particularly interesting type of nuclear reaction has been found by Chadwick and Goldhaber. They used radiation in the form of γ-rays, and showed that the deuteron then breaks up into a proton and a neutron, according to the formula

$$_1D^2 + \gamma \rightarrow {}_1H^1 + {}_0n^1.$$

The importance of this reaction lies in the fact that here the simplest composite nucleus is directly decomposed into its constituents. Besides, the reaction yields a very accurate value for the mass of the neutron, which is found to be practically identical with that of the proton.

8. Nuclear Structure.

With this material at our disposal, we shall now attempt to picture how the nuclei are built up.

The difficulty is that we know practically nothing about the forces which are acting in these very small regions. There is nothing for it but to proceed in the reverse way, inferring the forces and their laws from the observed facts.

When a nucleus flies to bits, whether spontaneously or on being broken down by the impact of a projectile, the fragments sometimes are themselves composite nuclei, such as helium nuclei or lithium nuclei, sometimes elementary particles. All four kinds of the latter occur: electrons, positrons, protons, neutrons. The simplest composite nucleus D breaks down into a neutron and a proton.

This suggests that perhaps all nuclei can be built up from neutrons and protons. As the addition of a neutron raises the mass by 1 but leaves the charge unaltered, whereas the addition of a proton raises both mass and charge by 1, every imaginable nucleus can be built up in this way. Helium, $_2He^4$, for example, would consist of two neutrons and two protons, the lithium isotope $_3Li^6$ of three neutrons and three protons, the lithium isotope $_3Li^7$ of four neutrons and three protons, and so on.

This idea is now generally accepted, for several reasons.

The older hypothesis that the nuclei consist of protons and electrons no longer has any justification now that the neutron and the positron have been discovered, and there are a number of facts that directly contradict it. For example, it can be shown experimentally that the nuclei rotate; they possess angular momentum. This follows from observations of various types. Many spectral lines exhibit a very

PLATE VIII

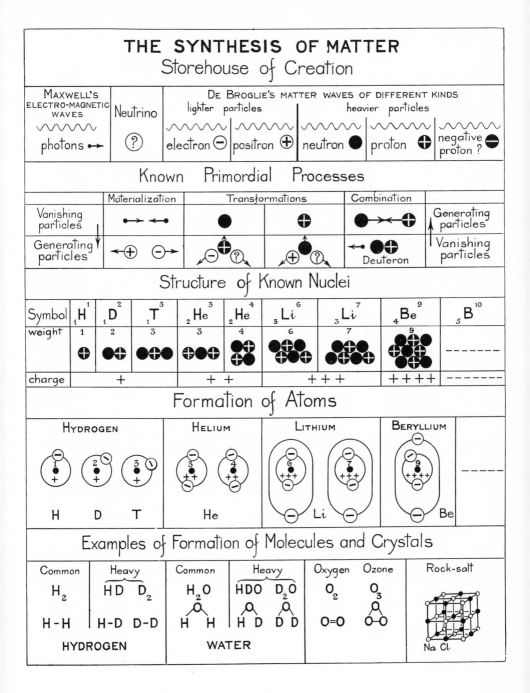

THE SYNTHESIS OF MATTER
Storehouse of Creation

fine structure, known as the *hyperfine structure*, which can be explained by the assumption that the nuclei execute quantized rotations, with quantization of direction (p. 184). From this the angular momentum of the nucleus can be inferred. Rabi has even succeeded in refining Stern's magnetic deviation method (p. 230) to such an extent that in many cases the mechanical and magnetic momenta of the nuclei can be detected and measured.

The chief result of this method and other less direct methods is that the proton has exactly the same " spin " as the electron, namely, $\frac{1}{2}$ (in quantum units, h being the unit of angular momentum). By the rules for the combination of spins, a nucleus should have an angular momentum of $\frac{1}{2}$, or $\frac{3}{2}$, or $\frac{5}{2}$, . . . if it consists of an odd number of particles (counting electrons and protons together), but 0, or 1, or 2, . . . if it consists of an even number of particles. This, however, is not the case. Nitrogen, $_7N^{14}$, for example, consists of 14 protons and 7 electrons, that is, 21 particles in all; this is an odd number, but the angular momentum is 1.

On the contrary, everything comes all right if we ascribe the spin $\frac{1}{2}$ to the neutron as confirmed by experiment. Then the nitrogen nucleus consists of 7 protons and 7 neutrons, that is, of 14 particles in all, a number compatible with the angular momentum 1.

Similarly in many other cases: the hypothesis of protons and electrons leads to contradictions, whereas the hypothesis of protons and neutrons is in agreement with experiment.

The assumption that the angular momentum of

the neutron is $\frac{1}{2}$ is in agreement with the experimental fact that the deuteron has the angular momentum 1; the spins of the proton and of the neutron, both equal to $\frac{1}{2}$, are simply added.

The most important argument for the hypothesis of protons and neutrons, however, is its power to explain a very striking regularity in the periodic system, namely, the fact that the atomic weights of the lighter atoms are often just double the atomic number, and that for heavier elements the atomic weight increases more rapidly than twice the atomic number. To do this, we need only make the following very plausible assumptions about the forces between neutrons and protons:

(1) neutrons have very little effect on one another;

(2) protons repel one another, owing to their charge;

(3) protons and neutrons attract one another with a very much stronger force, which, like chemical forces, exhibits the phenomenon of saturation.

True, we do not know the exact nature of these forces, but we can get a rough idea of how they arise. In the case of chemical forces the attraction is due to the exchange of electrons in the outer shells. We may imagine that a similar state of affairs holds for the proton and the neutron; the proton gives its excess charge up to the neutron, so that the latter becomes a proton, and then the process is reversed. We may even think of it as the direct exchange of a positron (or, if we start from the neutron, of an electron). However this may be, the effect of the force is certainly made immediately obvious by the existence of the deuteron.

Now the regularity which we mentioned above can be explained as follows, according to Heisenberg. As many saturated neutron-proton pairs are formed as possible. A packet of these pairs forms a nucleus, whose mass, of course, is double its charge. If, however, the number of particles becomes large, the electrical repulsion of the protons comes into play; the protons tend to force out other protons, so that a system with more neutrons than protons will be more stable; that is, the mass will be greater than twice the charge.

This argument can be extended and refined to cover a great many details.

Whence, then, do the electrons and positrons which occasionally fly out of the nucleus come?

Here again the hypothesis that the nuclei are built up of protons and neutrons has a great advantage. It reduces the emission of particles of light mass to a *single* elementary process; and there are good experimental grounds for this. For electrons and positrons never appear when the nucleus is instantaneously smashed by a particle hitting it; but they may appear in either natural or induced radioactive disintegrations.

We assume that the electrons and positrons arise from the transformations mentioned on p. 266:

$$\text{Proton} \rightarrow \text{Neutron} + \text{Positron},$$
$$\text{Neutron} \rightarrow \text{Proton} + \text{Electron}.$$

These will occur spontaneously if the nucleus thereby assumes a state of greater strength. Suppose, say, that by the escape of an α-particle a nucleus has been produced which has one neutron more than is good for it; then the neutron gives up an electron and changes into a proton, which is held more firmly.

A difficulty arises in connexion with the spins: neutron, proton, and electron each have the spin $\frac{1}{2}$; if the proton and the electron combine, the spin could be either $\frac{1}{2} - \frac{1}{2} = 0$ or $\frac{1}{2} + \frac{1}{2} = 1$, but not $\frac{1}{2}$. But worse is to follow!

These emissions of particles of light mass all exhibit a peculiarly sinister feature: the particles have no definite energy, but fly out with all possible velocities. Yet the nucleus left behind seems to be well defined. This is a very difficult point, a serious challenge to nuclear physics. If the nucleus is in a definite state before and after emission, it must have given up a definite quantity of energy. How, then, can the electrons shot out have varying energies? There are only two ways open; either the assumption that energy is created out of nothing or annihilated—which we feel very unwilling to adopt—or the assumption that there is a new invisible particle, not accessible to direct observation, which secretly carries away the missing energy. This particle would have to have the spin $\frac{1}{2}$; the difficulty about the combination of the spins, mentioned above, would then disappear. Further, this particle would have to be uncharged, and as light as an electron—if possible, it should, like a light-quantum, have no rest-mass at all. It has been given the fine-sounding name of "neutrino". On this basis Fermi has succeeded in developing a satisfactory theory of β-disintegration.

An unbiassed survey of the present state of nuclear physics, as summarized in Plate VIII (facing p. 272) reveals that we are only at the beginning of things. The printer's " devil " once played me the trick of changing " nuclear physics " into " unclear physics ". He was not far wrong. I am convinced that the dual conception of matter, as particles which act on one another by means of the electromagnetic field, cannot be final. Particle and field must form a higher unity; they must be much more intimately related to one another than is assumed in the wave mechanics.

The riddle of matter is still unsolved, but it is reduced to the problem of the ultimate particles. The solution of this problem is the task of the physics of the future.

Conclusion.

We have reached the end of our journey into the depths of matter. We have sought for firm ground and found none. The deeper we penetrate, the more restless becomes the universe, and the vaguer and cloudier. It is said that Archimedes, full of pride in his machines, cried, " Give me a place to stand, and I will move the world!" There is no fixed place in the universe: all is rushing about and vibrating in a wild dance. But not for that reason only is Archimedes' saying pontifical. To move the world would mean contravening its laws; but these are strict and invariable.

The scientist's urge to investigate, like the faith of the devout or the inspiration of the artist, is an expression of mankind's longing for something fixed, something at rest in the universal whirl: God, Beauty, Truth.

Truth is what the scientist aims at. He finds nothing at rest, nothing enduring, in the universe. Not everything is knowable, still less is predictable. But the mind of man is capable of grasping and understanding at least a part of Creation; amid the flight of phenomena stands the immutable pole of law.

So schaff' ich am sausenden Webstuhl der Zeit
Und wirke der Gottheit lebendiges Kleid.

Goethe, *Faust.*

'Tis thus at the roaring Loom of Time I ply,
And weave for God the Garment thou seest Him by.

Carlyle.

POSTSCRIPT

1. *Science and History.*

SINCE I wrote the last lines, 15 years ago, great and formidable events have happened. The dance of atoms, electrons, and nuclei, which in all its fury is subject to God's eternal laws, has been entangled with another restless universe which may well be the Devil's: the human struggle for power and domination, which eventually becomes history. My optimistic enthusiasm about the disinterested search for truth has been severely shaken. I wonder at my simplemindedness when I re-read what I said (p. 267) on the modern fulfillment of the alchemists' dream:

"Now however, the motive is not the lust for gold, cloaked by the mystery of magic arts, but the scientists' pure curiosity. For it is clear from the beginning that we may not expect wealth too."

Gold means power, power to rule and to have a big share in the riches of this world. Modern alchemy is even a short-cut to this end, it provides power directly; a power to dominate and to threaten and hurt on a scale never heard of before. And this power we have actually seen displayed in ruthless acts of warfare, in the devastation of whole cities

279

and the destruction of their population. Such acts, of course, have been achieved by other means. In the same war other cities than Hiroshima, with a considerable percentage of their population have been destroyed a little slower by ordinary explosives. Every previous war had its technical "progress" in destruction, back to the stone age when the first bronze weapons conquered flint axes and arrow heads. Still there is a difference. Many states, populations, civilizations have perished through superior power, but there were vast regions unaffected and room was left for new growth. To-day the globe has become small, and the human race is confronted with the possibility of final self-destruction.

When the question of a new edition of this book arose I felt a considerable embarrassment. To bring it up-to-date I had to write an account of the scientific development since 1935. But although this period is as full of fascinating discoveries, ideas, theories, as any previous epoch, I could not possibly describe them in the same tone in which the book was written; namely, in the belief that a deep insight of the workshop of nature was the first step towards a rational philosophy and to worldly wisdom. It seems to me that the scientists who led the way to the atomic bomb were extremely skillful and ingenious, but not wise men. They delivered the fruits of their discoveries unconditionally into the hands of politicians and soldiers; thus they lost their moral innocence and their intellectual freedom.

Therefore I have to apologize if the following account is short and dry. Its purpose is not to raise

PLATE IX

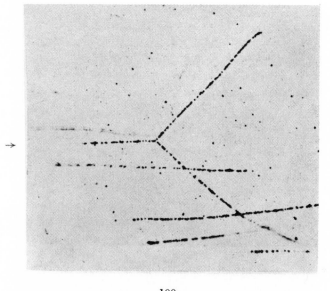

$\leftarrow -------- 100\,\mu --------- \rightarrow$

Scattering of protons by protons

(From Powell and Occhialini, *Nuclear Physics in Photographs* (Oxford))

enthusiasm, but to describe how the present tragic situation has arisen.

2. *Alterations and Corrections.*

I had not much to change of the original text of the book. Apart from numerous small improvements I have made only two major alterations. One (p. 88) concerns the self-energy of the electron. My hope that the non-linear theory of the electromagnetic field, which I had worked out (1933/34) in collaboration with some of my pupils, would settle this problem has not been fulfilled. It provides a satisfactory solution in the frame of the classical field theory; as shown by Schrödinger, the electromagnetic self-energy of a point singularity represents at the same time mass in all its aspects: as a measure of inertia and of gravitation. But the application of quantum theory to non-linear field equations proved to be most difficult and unsatisfactory. The reason, as pointed out by Heitler, is this: The electromagnetic field is characterised by two constants, the velocity of light c and the charge of the electron e, while quantum theory is characterised by Planck's constant h (or by $h = h/2\pi$). Now it turns out that the product hc has the same dimension as e^2, so that hc/e^2 is a dimensionless, numerical constant; its reciprocal is called the fine structure constant, as it determines, according to Sommerfeld, the splitting of the terms of the hydrogen atom (p. 181). The numerical value of this combination, $hc/e^2 = 137$, is not a small quantity. Now the classical theory is that limiting case of quantum theory where h is supposed to be small; for a given charge e the combination hc/e^2 ought to be small too, while in fact it is a rather large number, 137. Hence it has no meaning to develop a classical theory of the electron with its actual value of charge.

A very great amount of work has been devoted to formulating and solving the problem of the self-energy of the electron from the standpoint of quantum theory. Just as in the classical case (p. 87) the self-energy tends to infinity if the radius is taken to be zero, but in a very different way. Much ingenuity has been used in attempts to avoid disastrous consequences of this infinity, which enters into the calculation of many observable quantities, and considerable success has been achieved by developing rules for systematically eliminating infinite terms. The most remarkable feat of this refined quantum theory of the electromagnetic field was achieved independently by Schwinger in America and Tomonaga in Japan. Observations by Lamb and Retherford, namely had revealed that Dirac's relativistic theory of the electron (p. 209), which was considered to be the last word in theoretical physics, was not quite correct. However, the deviations could be explained by the authors mentioned. In spite of this success, the present theory of the electron and the electromagnetic field is not at all satisfactory. The "infinities" are not removed but shifted to a place where they are harmless for the time being.

The second major alteration (p. 162) concerns a question of terminology, but is important as it has to do with the philosophical foundations of quantum theory. In the original text I have used the expression "complementarity" to describe the two aspects of rays of light or matter: particles and waves. But that is not what Bohr had in mind, as I have learned from numerous discussions with him. He proposes to use this expression for two physical situations, produced by experimental arrangements, which have the same material object but are intended for the determination of different "conjugate" properties restricted by the uncertainty relation (p. 160). Such arrangements are mutually exclusive and complementary in this sense, that they define together all observable features of the object. The conceptions "particles" and "waves" have no such complementary character, as in many cases both are needed for a proper prediction of observations (the waves giving the probabilities of finding particles). One can speak here about the "dual" aspect of matter. This distinction seems to be only a question of words, yet it is essential—for what else is the meaning of natural philosophy but to give an account of natural phenomena in comprehensible terms? I think that Bohr's formulations, though often vague, are the best we have, and so I have changed the text.

3. Progress in Harmless Physics.

I shall now try to sketch the progress in atomic physics since 1935, and I propose to begin with the "harmless" kind of work which has nothing to do with super-explosives. I can mention only a few important points.

A powerful method for atomic research has been developed in the very short electromagnetic waves. The technique to produce these so-called "radar" waves was originally developed for war purposes—one can nowadays never get rid of this sinister side of research; but as radar is less a weapon than an aid to defence, we may forget about it. The importance of electromagnetic vibrations of high frequency for atomic research can be understood when we remember that the structure of the spectra of atoms is due to rotational motions, or precessions, of the electrons (Ch. IV) ; for instance, the Zeeman effect is due to the precession of the axis of rotation of the electrons or of the resultant vector angular momentum in a magnetic field. Instead of observing the energy differences of these rotational states as spectral lines, one can try to find the states themselves by exposing the system to an electromagnetic wave. If the frequency of this wave is very near, "in resonance" with, any of the atomic frequencies the propagation of the wave will be strongly affected (it will be absorbed or scattered), and thus the resonance frequency can be experimentally detected and measured. In this way a direct confirmation of the theoretical pictures of intra-atomic motions has been obtained and their frequency measured with a high degree of accuracy.

This consideration shows the importance of extending the range of frequency of electromagnetic waves over the whole spectrum. In fact the radar waves are almost completely bridging the gap between Hertzian waves, as used in broadcasting, and the longest infra-red waves, emitted from incandescent bodies. A new and powerful weapon to investigate atomic structures has thus been developed. One example is the method of Rabi, mentioned in the text (p. 230), to determine the angular momenta of nuclei. He uses, like Stern (p. 229), the deflection of atomic rays in an inhomogeneous field but superposes this by a vibrating field of adjustable frequency. If this is in resonance with one of the frequencies of precession of the nuclear angular momenta, the corresponding atoms are strongly affected and the atomic beam weakened. In this way the angular momenta and magnetic moments of many nuclei have been measured; in particular it has been confirmed that the spins of the proton and the neutron are both $\frac{1}{2}$.

Another example is the discovery by Lamb and Retherford, (mentioned already in the Post Script, p.282) of small deviations in the fine structure of the hydrogen atom from the predictions of Dirac's theory. These observations first made with the resonance method have been confirmed by Kuhn and Series with conventional spectroscopic methods.

A wide field of application of the new method is the magnetic behaviour of all kind of substances, where it has led to a vast extension of our knowledge and its theoretical explanation. But such details are outside the scope of this book.

The same holds for many other branches of atomic physics in which the last 15 years have brought great progress; for instance the theory of the chemical forces mentioned on p. 233. A description of these achievements would alone fill another volume, but as no essentially new ideas have appeared, it may be omitted from this sketch.

4. *Experimental Progress in Nuclear Physics.*

We now come to nuclear physics and shall begin with a short description of startling advancements in experimental technique.

There is first the counter-controlled cloud chamber, invented by Blackett and employed by him and others, mainly for the study of cosmic rays (p. 262–268). These extremely fast particles would be a most desirable addition to our armour for smashing nuclei if the chance of catching by a haphazard exposure a really interesting event, a head-on collision, would not be extremely small. To improve this situation Blackett combines the Wilson chamber with two counters, one above, and one below the chamber, connected in such a way that an electric impulse is only produced when a particle passes through both counters and hence also through the chamber. The current impulse is used to set the mechanism of the chamber in motion, including instantaneous illumination for taking a photograph of the track of the water droplets formed. By this ingenious gadget the number of collisions available for study has been immensely increased.

The same result has been obtained by quite another method, discovered by two lady physicists of Vienna, Miss Blau and Miss Wurmbacher. They discovered that photographic plates can be used for

PLATE X

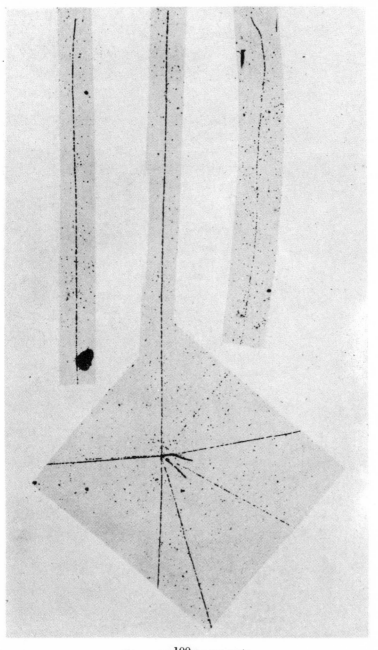

← − − − − − 100 μ − − − − − →

Tracks of particles of different types

Disintegration of a heavy nucleus by cosmic radiation, together with the tracks of a meson (top right) and a triton (top left). Note the marked difference in the scattering of the meson as compared with that of the heavy particles, and the rapid change in grain-density along the track. The long track from the explosion is due to an α-particle.

(From Powell and Occhialini, *Nuclear Physics in Photographs* (Oxford))

registering nuclear events. On ،a plate exposed to radio-active or cosmic radiation there appear tiny tracks, visible under a microscope. The bombarding particle obviously affects the grains of the emulsion in the same way as light. This is a very powerful method for nuclear research, compared with which the Wilson chamber appears as a clumsy instrument; for on a single square inch of a plate one finds often thousands of tracks. But before this method could be actually used much work had to be done by the manufacturers of photographic plates, who had to produce emulsions with much finer grain than usual, and by the physicists who had to learn to distinguish the tracks due to different kinds of particles and devise methods for studying the enormous amount of material thus obtained. The greatest contribution to this work has been made by the Bristol school of physics under Powell (Plate IX and X).

The progress in producing fast particles in great numbers is just as remarkable as that in observing them.

The "linear" accelerators, as used by Cockroft and Walton (p. 268) have been improved, by using higher voltages. An interesting machine which can be described as a gigantic magnification of the electrostatic induction apparatus of our school days has been developed by van der Graaff.

But the main progress in producing fast particles has been made by using relatively small voltages and repeated accelerations. In most of these instruments charged particles are driven around on circular or spiral orbits with the help of a magnetic field (p.

68). The acceleration can be applied in different ways. In the so-called Betatron of Kerst and Serber electrons are accelerated by the ordinary magnetic induction effect; over the constant magnetic field an alternating one is superposed which vibrates in the rhythm of the electronic revolutions. In the Cyclotron of Lawrence ions are driven around in such a way that by passing a certain diameter of the circular path they are electrically accelerated with the help of an oscillating system tuned to the correct rhythm. In the Synchroton of Veksler a combination of both principles has been used. A description of the technical details of these marvellous machines is quite beyond the frame of this book—and also beyond my own abilities.

In this way enormous particle energies have been reached. In terms of the kinetic energy of an electron accelerated by one volt electric potential, the self-energy mc^2 of the electron is about half a million. The voltages used in linear accelerators are some hundred thousand to a million volts. The circular orbit machines, on the other hand, produce particles which in these units surpass the hundred million mark and are approaching the thousand million. This value is a desirable aim, since it represents the order of magnitude of the self-energy of the proton and the neutron, which is 1836 times that of the electron (2000 times half a million is 1000 million).

As in cosmic rays still faster particles are known, there will be no end in this effort of building powerful accelerating machines. Thus nuclear physics,

PLATE XI

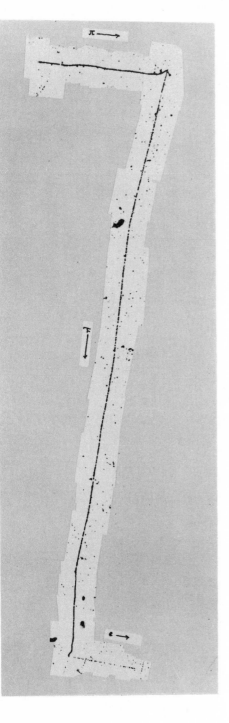

A primary π-meson decays into a μ-meson and an invisible (neutral) particle. The direction of the μ-meson is seen from the increase of the number of black grains. The μ-meson finally disintegrates into an electron and an invisible particle. The picture is a mosaic of microphotographs of short sections of the tracks taken with different focus.

which in the times of Rutherford was a game of wit
and skill, played by individuals with very small
means, has been transformed into an engineer's job,
performed by teams of specialists and costing millions of dollars.

5. *Progress in Nuclear Theory.*

What has been achieved with this enormous in-
vestment of ingenuity, organisation and money?
Again the volume and variety of the results defeat
all attempts to describe them in such a short account
as this. Innumerable nuclear reactions have been
observed, their output measured, new isotopes found
and studied, the characteristic constants of the particles derived, and so on. The knowledge of the
nuclei, their properties and reactions has immensely
increased. This is of great value to the specialist.
But in this book we have been not so much interested
in the accumulation of facts, but in the question:
What can we learn from these facts about the ultimate laws of nature, about the plan, the blue-print
of the Restless Universe?

If you turn back to p. 277 you will see that in
1935 the misprint "unclear physics" seemed to me
not a bad description of the state in which nuclear
physics then was. To-day this joke is out of date.
We have learned something about nuclear forces in
these 15 years, though we are still far from a full
understanding.

The progress of nuclear theory followed two
clearly distinct lines. It was natural first to apply
the well established laws of quantum mechanics and

to see how far one could get. It turned out that in this way many general features of nuclear physics could be understood—but not all. A step into the unknown had to be made and was made with considerable success. We begin with a short account of the first period.

6. *Nuclear Binding Energy and Stability.*

There are a number of properties of nuclei and their reactions which can be explained without a knowledge of the exact nature of the nuclear forces, solely from the fact that they are of short range and are depending on the spins and charges of the particles involved. This is due to the fact that the velocity of the heavy particles (protons, neutrons) is small, their de Broglie wave length therefore considerable, larger than the range of the forces. If quantum mechanics is applied to such a system it turns out that the stationary states, and also the effective cross-sections of collisions depend very little on how the force changes with the distance but much more on how it depends on the spins and the charges (and their exchange between the particles); but the latter type of forces is essentially determined by very general considerations of symmetry, without special assumptions (apart from a few constants). Thus a quite satisfactory theory of the deuteron and other simple nuclei has been developed.

For the more complicated nuclei even still cruder methods have led to good results. On p. 251 it was explained how the binding energy can be determined from the experimentally measured mass defect. The result is that the binding energy decreases with increasing mass number until a minimum is reached near the element iron, then it increases again. That means that all nuclei are in principle unstable, apart from those in the middle, near iron, and ought to transform themselves into stabler ones, the light ones by combination, "fusion," the heavy ones by disintegration, "fission," until the stable "iron" state is reached. This catastrophe is only prevented by the extremely small velocity of these reactions—yet there are ways to accelerate them, as the story of the atomic bomb has shown. Now all this can be understood by the most primitive model, suggested by Weizsäcker. One has only to assume that the addition of every nuclear particle (proton or neutron) liberates about the same amount of energy (of short range type) and then to calculate the total energy stored in the nucleus, taking into account that some of the particles lie on the surface and give therefore a smaller contribution than those in the interior. Thus the total energy of the short range forces has a term proportional to the volume and another proportional to the surface; to this one has to add the electrostatic energy of the protons (p. 275) and one obtains an expression of the binding energy in terms of the number of neutrons and protons which represents fairly well the region where actual nuclei can exist. Thus the general features of the stability of nuclei are explicable by very simple means. Subtler questions, for instance about the number of isotopes of a given element, need of course a more refined theory.

Niels Bohr used this model of a nucleus to explain many other nuclear phenomena which are of a dynamical character. It is suggestive to compare the state of nuclear matter to that of a solid or liquid piece of ordinary matter; but there is no indication of a nucleus having a solid crystal structure —in fact it cannot be expected because according to quantum mechanics even in the lowest state of a system there is a great amount of kinetic energy ("zero-point energy"). One speaks therefore of the "liquid drop (or droplet) model" of the nucleus. Now if such a droplet is hit by a bombarding particle, it is obvious that this is not simply deflected; it will penetrate into the interior and transfer its energy to the bulk of the nuclear particles, "heating up" the total mass. This heated nucleus will then evaporate just like an ordinary drop of water; that means it will emit one of its particles (or some of them). This is exactly what one observes: the bombardment of nuclei leads very rarely to simple scattering but to absorption of the primary particle with subsequent emission of another particle (artificial radio-activity). This simple idea has led to an elaborate theory of nuclear reactions which was quite successful.

7. *Mesons.*

All this is still on the first level of nuclear theory which uses well known laws and primitive pictures. The second level was reached by the Japanese Yukawa who opened the way to a deeper understanding of the nuclear forces. This fundamental progress was achieved not by detailed analysis of accumulated

experimental material but by the rational interpretation of its most general features. It was not the
result of collective enterprise and team work, but
the stroke of genius of a single man. It resembles
in this respect Maxwell's electromagnetic theory of
light; it is even methodically quite closely related to
this great discovery. The common concept is the
idea of a field of force. Inspired by Faraday's intuitive descriptions of electromagnetic interactions,
Maxwell (p. 81, 115) replaced the electrostatic
force acting instantaneously from charge to charge
by a field of force surrounding the charges, which is
determined by a law connecting its strength at a
point to that at infinitesimally near points, (a so-
called differential equation). If it is then assumed
that there is a little time lost by handing over energy
from point to point a disturbance of the field does
not proceed instantaneously but with a finite velocity
(as described in more detail on page 116), and one
has electromagnetic waves.

In the same way Yukawa constructed a short
range force in such a way that it could be translated
into the field language. It turned out that this can
be done in practically one way only. Of course the
differential equation is a little more involved than
the electrostatic one since it contains a certain constant a of the dimension length, representing the
range of the nuclear forces. Yukawa then imitated
Maxwell's second step, introducing the retardation
in time for the propagation of a disturbance. There
is no ambiguity in this step as the principle of relativity completely determines the dynamical law if

the statical law is given. Thus Yukawa was led to predict the existence of a new type of waves, nuclear force waves, and calculated their velocity which turned out to be smaller than that of light by an amount depending on the range constant *a*.

The question now was, can these waves be observed? Yukawa made another decisive step, by applying the principles of quantum theory: Just as an electromagnetic wave is connected with light quanta or photons, a wave of nuclear force must be connected with a new type of quanta. He calculated their rest-mass (p. 83) and found the simple formula

$$m = \frac{h}{ac}$$

If one introduces here the well known values of the quantum constant *h*, the velocity *c* of light and the range *a* of nuclear forces (some 10^{-13} cm) one obtains a value of 200 to 300 electron masses (one-sixth of the proton mass). Thus Yukawa's theory leads to the prediction of the existence of particles with masses "midway" between the electron and the proton; they were later accordingly christened "mesons" (Greek for middle).

Do they really exist? The first indication came from the cosmic rays; they have a penetrating component which prevails on sea level and gives tracks in the Wilson Chamber which do not look like those of electrons, but of heavier particles. Deflection in magnetic fields shows that their charge is just as often negative as positive, hence it is improbable that they are protons. Eventually the mass could be

PLATE XII

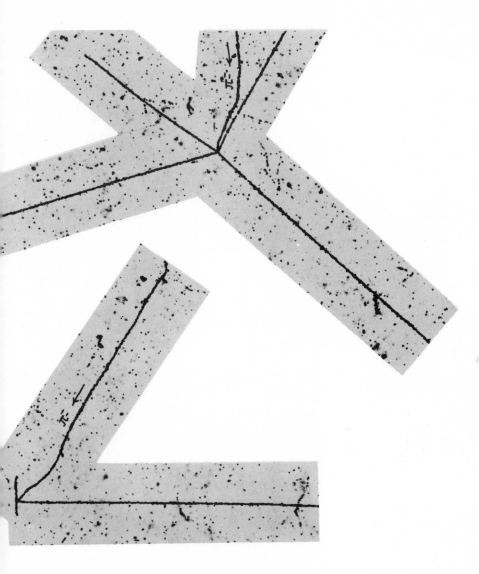

Nuclear disintegrations produced
by π-mesons

determined and was found to be about 200 times
that of the electron. This was the first experimental evidence for the existence of mesons. Then all
doubts were eventually removed by observing their
tracks on the photographic plate (Plate IX) and by
producing them artificially with the help of the cyclotron. But new complications arose. Yukawa himself
had already considered the question of why his new
particles do not form part of ordinary matter. His
answer was that they are not stable; after a very
short life they decay into electrons and neutrinos
(p. 276). That they decay has been confirmed by
direct experiments; for instance by comparing the
weakening of cosmic rays at sea level, largely com-
posed of mesons, by a sheet of condensed matter
(the earth above an underground station) with that
of a sheet of air containing the same number of
nuclei: in the air many more mesons are lost, which
shows that because of its larger extension more
mesons have time to decay. But when the photographic method was applied to the problem, the
tracks found revealed quite a new phenomenon: a
primary meson, emitted by a nucleus suddenly disappeared and another particle of somewhat smaller
mass was ejected. (There must be, of course, another invisible, i.e. neutral, particle recoiling to preserve balance of momentum).

Thus it was established that there are two kinds
of meson, a heavy type of about 300 electron masses,
called μ-mesons, and a lighter type of about 200
electron masses, called π-mesons. The μ-mesons
must be those contemplated by Yukawa, namely the

quanta belonging to the fields of nuclear forces; for they are emitted by nuclei and react violently with nuclei when they hit them, producing shattering explosions (Plate X). The π–mesons decay spontaneously into electrons and (presumably) neutrinos; they must be connected with the β–disintegration of nuclei (Plate X).

Yukawa's discovery has produced a flood of experimental and theoretical investigations and changed the whole situation in nuclear physics. There seems to be some evidence of other mesons; one of about 1000 electron masses has been well established. One thing seems certain: that the particles constituting ordinary matter, protons, neutrons and electrons, are only particularly stable examples in a multitude of possible particles with different intermediate masses. Thus the central problem of physics has shifted: instead of studying the laws of motion and of interaction of a few particles presented to us by nature, we have to find out why nature has just created these particles which exist. Or expressed in a less mystical way: We have to look for a general principle from which the possible fields and the corresponding particles can be derived, together with the numerical ratios of their masses and their coupling constants (by which I mean quantities like the electric charge which couples electrons and photons). Attempts in this direction begin to emerge; I myself have made one, under the title "Principle of Reciprocity" which produces many particles with definite mass ratios—only too many, an infinite number for which there seems to be no justification in experience.

However primitive these first trials may be, it seems very likely that Yukawa's work was a turning point in the search for the ultimate laws of nature.

8. *Nuclear Fission.*

The reader may find that my account of these events reflects some of the same enthusiasm with which the old book was written. Indeed, they show that the spirit of research, the disinterested desire to disclose nature's mysteries, is as lively and active to-day as it ever was, and even more so as it has spread from a few centres of learning in Europe and America to the whole world and all nations. But it is just this fascination of lifting the veil of mystery and of discovering harmony in apparent chaos which caused the present crisis. The satisfaction of the noble curiosity of the scholar is only one aspect of research. Science is also—and many say predominantly—a collective effort to obtain power over the forces of nature in the interest of human life. That is the root of the trouble.

The story begins harmlessly enough. Physicists were experimenting with the heaviest nuclei, thorium and uranium, in order to find out whether, by capturing neutrons, these might not be turned into "transuranium" elements, with an atomic number higher than 92. Two Germans, Hahn and Strassmann, however discovered that neutron bombardment had quite a different effect: the uranium nucleus broke up into two parts of comparable size, for instance

$$_{92}U \rightarrow {}_{56}Ba + {}_{36}Kr.$$

Processes of this kind which are accompanied by the liberation of an enormous amount of energy, have been called "fission." Two other Germans, Frisch and Lise Meitner, suggested presently a simple explanation on the lines of Bohr's droplet model, which we have just discussed. The Th and U nuclei contain so many protons that the repulsive energy of the electrostatic forces is almost as great as that of the short-range nuclear (meson) forces; if a neutron is captured the stability limit is surpassed, and the nucleus disintegrates into fragments of about equal size which fly apart with a surplus energy 8 to 10 times as large as that set free in ordinary nuclear reactions. Bohr and Wheeler investigated this instability in more detail and were able to predict the efficiency (the "cross-section" of effective collisions) of neutrons with varying velocity. One of their results opened possibilities of immense importance.

The process of fission of uranium is accompanied by the emission of neutrons which are either simultaneous with the breaking up, or appear with a time lag as disintegration products of the unstable fragments. Now each of these neutrons may hit another uranium nucleus and induce it to fission, neutrons being again emitted; and so on: an avalanche of fission may be started. Such "chain reactions" are well known in ordinary chemistry. If it were possible to produce a chain of fission the enormous energy accumulated in the heavy nuclei could be utilized for practical purposes. Observations of the efficiency of neutrons of different velocity in producing fission showed that fast neutrons had about the same effect on thorium and uranium, while slow neutrons were much more efficient in uranium. Bohr deduced from his theoretical considerations that the ordinary uranuim isotope $_{92}U^{238}$ could not account for this fact and suggested that it was due to the presence of a small amount (o.7%) of the isotope $_{92}U^{235}$. This prediction was later completely confirmed by Lawrence and his collaborators.

To this point it was possible to give an account of scientific progress as if it had no connection with anything else which happened in the world. This was the attitude of most scientists of the older generation to which I belong, and my book is written in this style. The civilised, peaceful and apparently stable society of Europe and America could afford the luxury of such abstractions.

9. *Political Fission and Nuclear Fusion.*

But in 1933 this period came to an end with Hitler's accession to power. A cleavage developed between his totalitarian system and the rest of the world where liberal, democratic and socialistic ideas still prevailed. It was an intellectual, ethical, political fission resulting in the ejection of brains from Germany which were absorbed in other communities. Among the refugees reaching the United States were excellent scientists. Germany had been, for about a hundred years, in the front line of learning and research. Theoretical physics in particular was on a high level; relativity and quantum theory both started in Germany, and many other important ideas as well, as can be seen by scanning the names quoted in this book. Thus America benefited from the expulsion of German scholars, and one case must be mentioned here. It concerns the explanation of the stellar energy with the help of nuclear fusion.

That this and other cosmological problems have not been discussed in this book is certainly regrettable. The word Universe in fact suggests to the mind more readily the stellar macrocosmos than the atomic microcosmos, and my only apology is the fact that I am a physicist, not an astronomer. Still these two worlds of the largest and the smallest dimensions are interrelated in the most intimate manner. All our knowledge of the stars comes from their emission of radiation, which is an atomic process and must be interpreted in terms of atomic theory.

Life on earth depends on the sun, a central god in all ancient mythologies: Ra in Egypt, Apollon in Hellas. An enormous amount of energy is permanently pouring out from the sun; where does it come from? The first reasonable answer to this question was given by Lord Kelvin, about a hundred years ago. He remarked that the sun must contract under its own weight, and calculated the heat produced by this process of conversion from gravitational energy. Though it is a very great amount it turned out not to be sufficient at all for the existence of the sun through the ages. In particular, when one learned to estimate the age of minerals in natural rocks on the earth by radioactive methods, the time scale derived from Kelvin's idea turned out much too small. These methods are easily understood: We know the decay period of radioactive substances and can there-

fore calculate the ratios of the quantities of the original element and of their disintegration products in
a given time. The other way round, an analysis of
these ratios in the interior of rocks allows the determination of the time which has elapsed since its
formation as a solid. Thus reliable values for the
age of the earth's crust have been obtained, and this,
as said before, was irreconcilable with the gravitational theory of the sun's energy.

But radioactivity also provided the clue to the
solution of this difficulty. We know that the disintegration of nuclei produces enormous amounts of
energy. Hence it seems obvious that the radiation
of the sun and the stars can be explained by assuming some nuclear processes in their interior. But
what are these processes?

The answer was given by one of the German refugee scholars in U.S.A., Hans Bethe. Most of the
stars are essentially balls of hydrogen, as can be
derived from their spectra and their densities (as
far as these are known). We have seen (p.291 of
this Postscript) that the light nuclei are in principle
unstable and have the tendency to unite to heavier
nuclei, by "fusion". On p. 251 we have calculated
the energy liberated by forming a helium nucleus
from four hydrogen nuclei (actually 2 protons and
2 neutrons) and found that it is bigger than for any
other combination of nuclei. This process would
therefore supply an ample amount of energy if it
could go on with a reasonable velocity. But that is
just the difficulty. As nuclei are positively charged

and therefore repel one another fusion needs high relative motion, i.e., high temperature. Moreover, a simultaneous collision of not less than four hydrogen nuclei to form a helium nucleus is obviously very improbable. The way out suggested by Bethe consists in what chemists call a catalytic process. The building up of helium does not happen in one act, but in a chain of four acts, each of which is a simple collision with another particle. As the first of these is one of the carbon isotopes one speaks of the carbon cycle. The catalysing nuclei are all reconstituted and the net result is the disappearance of four hydrogen nuclei and the formation of a helium nucleus. This theory was so successful in explaining the heat generated in the sun and the stars that the existence of fusion could hardly be doubted.

Thus it was clear that ordinary matter is unstable, explosive, that we are sitting on a powder barrel. Still danger seemed remote as ordinary sparks would not work; temperatures of million degrees, as in the centre of stars, are needed for ignition. Fission on the other hand would need no ignition through high temperatures, because a neutron is not repelled by the nuclear charge. All depends on whether a chain of fission could be established.

10. *The Atomic Bomb.*

Nobody can say how this problem which was in the mind of many physicists would have developed if the war had not interfered. Very likely the progress would have been much slower, possibly slow enough for a contemplation of the consequences, economic and political. As it wàs, a kind of panic developed. The refugee scholars spread not only knowledge, but also the conviction that the Hitler Government would do its utmost to obtain a nuclear explosive and would not hesitate to use it ruthlessly. British and American physicists were soon persuaded that the possibilities of fission had to be investigated, and many laboratories, many teams of theoreticians began to work.

I have not been a member of one of these groups, and I know about their work not more than anybody may grasp from the publications. Therefore I cannot make any revelations or add personal observations to the facts generally known. I shall only repeat the outline of events.

It was found that a chain of fission could be expected provided that the loss of neutrons is kept low enough; this depends mainly on two conditions: avoiding of absorption by keeping the uranium extremely pure, and avoiding escape by using a block of sufficient size. The technical problems set by these demands are formidable, and so are the expenses which the U.S.A. Government allocated to the project. The problem consisted in extracting the rare uranium isotope U^{235} from the natural metal. All known processes for separating chemically identical atoms of nearly equal mass are extremely inefficient and slow. However by using enormous apparatus, in which processes like diffusion were applied in many repetitions, the concentration of U^{235} was considerably increased, and the final separation was achieved with the help of electromagnetic separation, by a kind of gigantic mass spectrograph (p. 247).

Another possibility of producing fissionable material was to build up transuranium elements. The nucleus $_{92}U^{239}$ formed when a neutron is captured by the ordinary uranium nucleus $_{92}U^{238}$ is likely to be unstable and to disintegrate by successive emissions of electrons into new elements, later called neptunium $_{93}Np^{239}$ and plutonium $_{94}Pu^{239}$. It seemed probable that plutonium 239 would react on the capture of a neutron by fission. These predictions have been confirmed by experiment. The method used in these experiments as well as in the final process of production of plutonium consisted in the building of a "reactor" or "pile", a regular arrangement of lumps or bars of uranium separated by a substance of light atoms (graphite or heavy water) which serves as "moderator" by reducing the velocity of the neutrons emitted. Fast neutrons of a certain range of speed produce plutonium from U^{238} while those slowed down to thermal velocity keep up the chain of fission of U^{235}. The reaction velocity can be controlled with the help of bars of absorbing material (cadmium).

The first self-contained chain reaction was established, under the direction of Fermi, on 2nd December 1942. The plutonium thus produced turned out to have the properties expected. As it differs chemically from uranium, it can be extracted by chemical means, which are simpler, faster and more efficient than the methods for the separation of isotopes.

Still the technical task is formidable since the whole process must be made in automatic factories with remote control because of the deadly radiation emitted by the material.

After the fissionable substances were available in sufficient amount the problem of the construction of a bomb had to be solved. There is a critical size for fission: a lump of uranium 235 or plutonium smaller than this size will be harmless, but one larger than this size will immediately explode, since there are always stray neutrons which initiate a chain of fission. Hence the problem was to bring two lumps together in the shortest possible time so that the best use of the material might be made.

The methods developed for this purpose are secret and of no great interest from the standpoint of science. They were completely successful. On July 16, 1945, the first experimental bomb was exploded near Los Alamos, New Mexico. This was certainly one of the greatest triumphs of theoretical physics if measured not by the subtlety of ideas but by the effort made in money, scientific collaboration and industrial organisation. No preliminary experiment was possible, the tremendous risk was taken in the confidence that the theoretical calculations based on laboratory experiments were accurate. Therefore it is no wonder that the physicists who watched the terrific phenomenon of the first nuclear explosion felt proud and relieved from a heavy responsibility. They had done a great service to their country and to the community of allied nations.

But, when a few weeks later, two "atomic bombs" were dropped over Japan and destroyed the crowded cities of Hiroshima and Nagasaki, they discovered that a more fundamental responsibility was on their shoulders.

11. *Tragic Fusion of Physics and Politics.*

The world had become pretty callous against the horrors of the war. Hitler's seed had grown. His was the idea of total war, and his bombs smashed Rotterdam and Coventry. But he found keen pupils. In the end the bombers of both sides succeeded in a systematic devastation of Central Europe. A great part of its historic and artistic treasures, the inheritance of thousands of years, went up in flames. An architectural jewel like Dresden was destroyed in one of the last days of the European war, and 20,000 civilians, men, women and children are said to have perished with it. I do not doubt that those responsible for this act can rightfully claim tactical and strategical necessity; and the world in general found sufficient justifications, ranging from blind hatred and the wish of retribution to the quasi-humane idea that to shorten the war all means are good enough. Ethical standards had fallen sharply, indeed.

Still the two atomic bombs dropped on Japan made a stir, and when details of the human tragedy became known there was something like an awakening of conscience in many parts of the world.

This is not the place to express my personal judgment of the statesmen who decided to use this brutal application of power. Cases of precedence are plen-

tiful—there is not much difference in the responsibility for killing 20,000 in one night or 50,000 in
one minute. But being a scientist I am concerned
with the question of how far science and scientists
share the responsibility.

The motives of those who took part in the development of nuclear explosives were certainly above
reproach: Many of them were just drafted to this
work as their war service, others joined it, driven
by the apprehension that the Germans might produce the bomb first. Yet there was no organisation
of scientists which could form a general opinion.
Single men became little cog-wheels in the tremendous machine, which was directed by political and
military authorities. The leading physicists became
scientific advisors of these authorities and experi-
enced the new sensation of power and influence. They
enjoyed their work and its tremendous success, and
forgot for the time being to think hard about its
consequences. It is true that a group of scientists
warned the U.S. Government not to use the bomb
against cities, but to demonstrate its existence and
power in a less murderous way, for instance on the
top of Fujiyama mountain. They predicted very accurately the disastrous political consequences which
an attack on a city would have. But their advice was
neglected.

The principal discrepancy between public opinion
in the United States and the conviction of the scientists is concerned with secrecy. The scientists are convinced that there is no secret in science. There may
be technical tricks which can be kept secret for a

limited period. But the laws of nature are open to anybody who is trained in using the scientific method of research.

Therefore it was futile to keep the atomic bomb project from being known to the Russian allies, and the maintenance of this secret has with necessity transformed them from friends into enemies. They felt menaced by a tremendous new weapon; they started to develop it themselves, and they obtained it in a shorter time than was ever expected.

On the other hand this phantom of secrecy had disastrous effects on the development of nuclear physics in America. Many physicists have been subjected to suspicion and even to accusation of disloyalty. The whole of science has been hampered by the classification of discoveries into secret and open ones, and by the supervision of publication. There is no doubt that certain security measures, mainly in regard to technical questions are unavoidable. But the subordination of fundamental research to political and military authorities is detrimental. The scientists themselves have learned by now that the period of unrestricted individualism in research has come to an end. They know that even the most abstract and remote ideas may one day become of great practical importance—like Einstein's law of equivalence of mass and energy. They have begun to organise themselves and to discuss the problem of their responsibility to human society. It should be left to these organisations to find a way to harmonise the security of the nations with the freedom of research and publication without which science must stagnate.

12. *Paradise or Hell.*

The release of nuclear energy is an event comparable to the first fire kindled by prehistoric man—though there is no modern Prometheus but teams of clever yet less heroic fellows, useless as inspiration for epic poetry. Many believe that the new discoveries may lead either to immense progress or to equal catastrophe, to paradise or to hell. I, however, think that this earth will remain what it always was: a mixture of heaven and hell, a battlefield of angels and devils. Let us have a look around: what are the prospects of this battle, and what can we do to help the good cause?

To begin with the devil's part, there is the hydrogen bomb. We have seen that, though almost all matter is unstable in principle, we are protected against nuclear catastrophe by the low temperatures on earth, which even in our hottest furnaces are quite insufficient to initiate nuclear fusion. But the discovery of fission has destroyed this security. The temperature in an exploding uranium bomb is presumably high enough to start the fusion of hydrogen with the help of the "carbon cycle," which is the source of stellar energy, or a similar catalytic process. Thus an explosive of many thousand times higher efficiency than the fission bomb could be made from a material available in abundance. Of course, work has started with the usual argumentation: if we do not do it, the other fellow (meaning the Russian) will. If it succeeds there will be a new instrument of wholesale destruction, but no peaceful application of the new forces seems to be possible. No way is known to slow down fusion in order to use it as a fuel. A perfectly hellish prospect.

Fission however has many and far reaching applications of a peaceful kind. It can be used as fuel, since the reaction velocity can be controlled. Each pile produces an enormous amount of heat which at present is wasted in most cases. Power stations using uranium or thorium as fuel are possible, as the difficulties connected with the pernicious radiation could certainly be overcome. The question is however an economic one. The raw material is rare, and if the same amount of energy which is at present made from coal would be produced by nuclear reactors, the whole uranium ores at present or in future available would be used up in less than half a century. Hence it is improbable that the new fuel will be able to compete with coal and oil. Under certain conditions however, this may be the case, namely where the advantage of the small bulk and weight of nuclear fuel, as compared with that of coal or oil, is decisive. There is a possibility of increasing the efficiency of fission by "breeding", i.e., by directing the process in a pile in such a way that a great proportion of the nuclei present is transformed in fissionable isotopes. This would mean an extension of the raw material over a much longer period.

Apart from the still problematic application of nuclear reactions for power production, there are numerous others which have already led to great progress and which are more promising. There is first the generation of new isotopes in the pile. Our knowledge of the stability of nuclei and of the laws of their interaction has been immensely increased. Some of the radio-active products can be used in medicine for therapeutical purposes, replacing for instance radium in the fight against cancer. The most important application is the so-called "tracer method" which is revolutionizing chemistry and biology. Already in the first period of radio-activity v. Hevesy had the idea to trace the fate of atoms in chemical or biological processes by adding to them a small amount of a radio-active isotope. This discloses its presence by radiation, and as the methods of detection of radiation are extremely sensitive, one can thus determine much smaller amounts of an element than with the balance. It is even possible to investigate the distribution of atoms in living tissue. The actual application of this idea was formerly restricted to the few atomic types for which naturally radio-active isotopes were known. Such isotopes are available for almost all elements of the periodic system. The work on this line, though hardly begun, has already led to important results, and will lead to still more.

13. *Outlook.*

But what are these important results compared with the spectre lurking in the background, the possibility of atomic warfare on a great scale?

In combination with other infernal contraptions, like rockets to deliver bombs at large distances, chemical, biological and radio-active poisons, such a war

must mean a degree of human suffering and degra-
dation which is beyond the power of imagination.
No country would be immune, but those with highly
developed industry would suffer most. It is very
doubtful whether our technological civilisation would
survive such a catastrophe. One may be inclined to
regard this as no great loss, but as a just punishment
for its shortcomings and sins: the lack of productive
genius in art and literature, the neglect of the moral
teachings of religion and philosophy, the slowness to
abandon outdated political conceptions, like national
sovereignty. Yet we are all involved in this tragedy,
and the instinct of self-preservation, the love of our
children, makes us think about a way of salvation.

There are the two political colossi, U.S.A. and
U.S.S.R., both pretending to aim at nothing but
peace, but both arming with all their power to de-
fend their ideology and way of life, and between
them is a weak and divided Europe, trying to steer
a middle course. Both sides are greedily devouring
the latest achievements of scientific technology for
their armed forces. Both have some kind of theory
for their way of life in which they believe with an
amazing fanaticism. Yet the foundations of these
theories are rather doubtful. They use the same
words for different or even opposite ideas, as for
instance "democracy," which in the West means a
system of parliamentary representation freely and
secretly elected, but in the East means something
quite different and hard to formulate (a complicated
economic and political pyramid of burocracy which
aims at representing, and working for "the people").

In other ways the American theory is much vaguer than the Russian, and that seems to have a historical reason. America has grown by expansion in a practical vacuum; the pioneers of the West had to overcome terrific natural obstacles, but negligible human resistance. The Russia of to-day had to conquer not only natural but human difficulties: she had to break up the rotten system of the Czars and to assimilate backward Asiatic tribes; now she has set herself the task of bringing her brand of modernisation to the ancient civilizations of the Far East. For this purpose it is indispensable to have a well defined doctrine full of slogans, which appeals to the needs and instincts of the poverty stricken masses. Thus one understands the power which Marx's philosophy has gained in the East. What can we scientists do in this conflict? We can join the spiritual, religious, philosophical forces, which reject war on ethical grounds. We can even attack the ideological foundations of the conflict itself. For science is not only the basis of technology but also the material for a sound philosophy. And the development of modern physics has enriched our thinking by a new principle of fundamental importance, the idea of complementarity. The fact that in an exact science like physics there are found mutually exclusive and complementary situations which cannot be described by the same concepts but need two kinds of expressions, can be applied to other fields of human activity and thought. We have mentioned (p. 165) some such applications to biology and psychology, suggested by Niels Bohr. In philosophy there is the ancient and central problem of free will. Any act of willing can be regarded on the one side as a spontaneous process in the conscious mind, on the other as a product of motives depending on past or present impressions from the outside world. If one assumes that the latter are subject to deterministic laws of nature, one has a conflict between the feeling of freedom of action and the necessity of a natural process. But if one regards this as an example of complementarity the apparent contradiction turns out to be nothing but an epistemological error. This is a healthy way of thinking, which properly applied may remove many violent disputes not only in philosophy but in all ways of life; for instance in politics.

Marxian philosophy, which is a hundred years old, knows of course nothing of this new principle. However, a prominent Russian scientist has recently attempted to interpret it from the standpoint of "dialectic materialism," which teaches that all thinking consists of a thesis opposed by an antithesis; after some struggle, they are combined in a synthesis. In this Marxian dogma, so he claims, you have the prediction of what has happened in physics, for instance in optics: Newton's thesis that light consists of particles was opposed by Huygens' antithesis that it consists of waves, until both were united in the synthesis of quantum mechanics. That is all very well and indisputable, though a little trivial. But why not go further and apply it to the two competing ideologies: Liberalism (or Capitalism) and Communism, as thesis and anti-thesis? Then one would expect a synthesis of some kind, instead of the Marxian doctrine of the complete and permanent victory of communism. It can hardly be expected that the ideas of Marx, developed about 100 years ago, can throw much light on the development of modern science. The opposite is more likely: that the new philosophical ideas developed by science during these 100 years may help towards a deeper understanding of social and political relations. Indeed, we find two systems of thought which deal with the same structure, the state, in completely different, apparently contradictory ways. One starts from the freedom of the individual as the basic conception, the other from the collective interest of the community.

This distinction corresponds roughly to the two aspects of the problem of willing which we have just mentioned: the subjective feeling of freedom on the one hand, the causal chain of motives on the other. Thus the West idealises political and economical liberalism, the East collective life regulated by an all-powerful state. But as it seems likely that the contradiction in the problem of free will can be solved by applying the idea of complementarity, the same will hold for the contradiction of political ideologies. Thus the intellectual gulf between West and East may be bridged, and that is the service which natural philosophy can offer in the present crisis.

The world which is so ready to use the gifts of science for mass destruction would do well to listen to this message of reconciliation and co-operation.